Facts Not Feelings

Mastering the Art of Running Your Marketing Agency
Based on Data and Not Emotions

By Chris Martinez

Facts Not Feelings
Mastering the Art of Running Your Marketing Agency Based on Data and Not
Emotions
Copyright © Website In 5 Days LLC

ISBN: [979-8-218-47645-8]

First Edition

Printed in the United States

Dedicated to my wife.

Thank you for putting up with the late nights, the mood swings, the smelly gis, and the uncertainty of entrepreneurship. I couldn't do this life without you.

Table of Contents

Introduction: **9**
Why Data Matters

Chapter 1: The Wake-Up Call   **25**

Chapter 2: Meeting the Mentor   **49**

Chapter 3: The Launch   **95**
Level 1

Chapter 4: Validation Stage   **123**
Level 2

Chapter 5: Growing Pains   **151**
Level 3

Chapter 6: The Professional Shift   **179**
Level 4

Chapter 7: The Crossroads   **211**
Level 5

Chapter 8: Lessons Learned   **227**
A Roadmap to Success

Facts Not Feelings

Introduction:
Why Data Matters

When I first started my outsourcing business, I believed in the power of intuition. After all, isn't that what entrepreneurship is all about? The ability to trust your gut, take bold risks, and navigate the uncertain waters of business? This mindset served me well in the beginning. We grew rapidly, and within a few years, our revenue soared to $800,000. It was a thrilling ride, filled with creative breakthroughs and exciting client projects. Yet, despite our impressive top-line growth, the bottom line told a different story. Our profits were marginal, and the financial health of the business was constantly in a precarious state.

I often found myself looking at other successful agencies and wondering what they knew that I didn't. I was convinced there was some secret sauce—some hidden knowledge that these industry leaders possessed. My late-night musings frequently led me to the same place: frustration and doubt. What was I missing? Why couldn't we break through to true financial stability and success?

The turning point came during a particularly challenging period. We had just lost several major clients due to a series of operational mishaps, and the financial strain was more apparent than ever. I was at a crossroads, desperate for answers and a way to turn things around.

With little money in the bank, I decided to invest in a training program to learn how to run the business more effectively. I was desperate and knew that if I didn't do something drastic, and by drastic I mean putting a $14k training program on a credit card, that I would always be stuck chasing profit.

Long story short this training program gave me exactly the insight that I needed. It wasn't that I was stupid or that I lacked the charisma to build a profitable business. In this short 12-week training I realized that I was running my business based on gut instinct alone and while intuition had gotten us off the ground, it was no longer enough to sustain and grow the agency. I needed to run the business like a professional, and that meant grounding my decisions in hard data.

This realization was both humbling and enlightening. I started to dig into the basics of business metrics—things I had previously considered boring or irrelevant. Terms like Gross Margins, EBITDA (Earnings Before Interest, Taxes, Depreciation, and Amortization), and organizational charts suddenly became my new focus. It dawned on me that these "boring" things were what successful business leaders knew and leveraged to their advantage. These metrics and structures were the foundation of sustainable growth and profitability.

Understanding Gross Margins was a revelation. I learned that it wasn't just about how much money we made, but how much we kept after covering the costs of delivering our services. This insight led me to scrutinize our pricing strategies and cost management practices. I discovered that while we were great at generating revenue, our cost structure was eating away at our profits. Adjusting our pricing models and cutting unnecessary expenses started to have an immediate positive impact on our margins.

EBITDA, another seemingly mundane concept, became a crucial indicator of our financial health. It provided a clearer picture of our operational performance by excluding the effects of financing and accounting decisions. This metric helped me see the true profitability of our core business activities. It was a wake-up call to focus on improving operational efficiency and reducing overhead costs.

Another critical area was the organizational structure. In our early days, roles were fluid, and responsibilities often overlapped. This flexibility worked when we were a small team, but as we grew, it led to confusion and inefficiencies. Learning about organizational charts helped me see the importance of clear roles and responsibilities. By defining these more precisely, we improved accountability and streamlined our workflows. Everyone knew what was expected of them and who to turn to for specific issues, reducing bottlenecks and enhancing productivity.

Data didn't just transform our financial and operational strategies; it also revolutionized how we approached marketing and client projects. By leveraging analytics, we gained deeper insights into market trends, client behaviors, and campaign performances. These insights allowed us to tailor our strategies more effectively, predict outcomes more accurately, and deliver better results for our clients.

For instance, instead of relying on gut feelings to shape our customer experience and retention strategies, we began to use data to drive our decisions. We analyzed which types of deliverables were driving the majority of our cancellations, identified the most effective communication channels created strong relationships, and optimized our onboarding to ensure we managed expectations. This shift not only improved the success rates of our deliverables but also strengthened our relationships with clients. They saw the value of our data-driven approach and trusted us more as strategic partners.

Transitioning to a data-driven business model wasn't without its challenges. It required a cultural shift within the agency, as everyone needed to embrace the importance of data and analytics. We invested in training our team, implementing new tools, and fostering a mindset that valued evidence-based decision-making. This transformation was gradual, but the results were undeniable. We became more agile, more efficient, and more profitable.

Reflecting on this journey, I realized that the gap between where we were and where we wanted to be was bridged by data. The successful people I had envied didn't have some mystical insight; they understood the power of numbers. They knew how to interpret data, derive actionable insights, and apply them to drive business decisions. This understanding was their true advantage, and it became ours too.

This book is a culmination of my journey from relying on gut instincts to embracing data-driven decision-making, but written through the eyes of a struggling agency owner named "Alex". It's a guide for entrepreneurs and business leaders who want to build sustainable, profitable, and scalable businesses. By sharing my experiences, insights, and the tools we used, I hope to help you navigate your own path to success. Running a marketing agency based on data is not just about improving your bottom line; it's about building a foundation that can support your vision and drive your growth. Welcome to the new era of professional, data-driven business management.

The Pitfalls of Gut Instinct

Picture this: a bunch of kids in a field, tossing a football around. It's chaotic, full of laughter, and improvised plays. That's your agency in its infancy—full of potential but all over the place. There's a certain charm to this disarray, a sense of freedom and possibility. Every play is a surprise, every outcome is unpredictable. It's pure, unadulterated joy. But as you well know, when it comes to delivering consistent results and growing a business, chaos is not your friend. This is where the journey to professionalization begins.

Now, catapult yourself to the NFL—the big leagues. It's

a different world here. The stadiums are massive, the stakes are higher, and the audience is vast. It's not just about talent; it's about precision, strategy, and flawless execution. Each player is not just good; they're the best in their position, and they've trained relentlessly to get there. They know their role like the back of their hand, and they perform it with a mix of rigorous discipline and creative flair. The coach, with years of experience and a keen eye for the game, sets the vision. The team executes it, and the operations crew, the unsung heroes, ensure everything is in place for the magic to happen.

But how do you go from that playful scrimmage in the field to the structured, strategic plays of the NFL? It's a journey of transformation, of learning to balance the free-spirited creativity of your agency's early days with the disciplined precision required to play in the big leagues. It's about instilling discipline without dampening creativity, about building processes that don't stifle but rather amplify your brilliance.

This is where the analogy to "Moneyball" comes into play. In professional sports, and increasingly in business, success is no longer about guessing or relying on instinct. It's about leveraging data and analytics to inform every decision. Just as investors analyze balance sheets, business leaders analyze performance metrics, market data, and financial reports to guide their strategies. The leadership team, like the coaches in the NFL, must know exactly what each team member needs to do to win games and get to the playoffs and the Super Bowl. It's about understanding and utilizing the science behind winning.

In the NFL, leadership is not just about inspiring the players to win. True leadership encompasses knowing the key performance indicators (KPIs) that each player needs to hit for the team to be successful. This is why players are given bonuses for scoring a certain number of touchdowns in a season or sacking the quarterback a specific number of times. These KPIs are derived from meticulous analysis of data, understanding the impact of each player's performance on the overall success of the team. Coaches and managers use this data to set clear, measurable goals and ensure they have the right players who can execute the plan and achieve these targets. They then hold the players accountable to these standards, driving performance and success.

Your agency is like a top NFL team. Your projects are your games, each with its own challenges and opportunities. Your clients are your fans, cheering you on and invested in your success. Your team are your employees, each member playing a crucial role. And just like in the NFL, the teams that win are the ones with not just the best players, but the best strategy, the best preparation, and the best execution.

In the NFL, every move is calculated, every play is practiced, and every role is clear. The quarterback knows the plays, the receivers know their routes, and the defense knows how to react to the opposing team's movements. In your agency, the same level of clarity and coordination is required. Your strategists plan the campaign, your creatives bring it to life, your account managers liaise with the clients, and your operations team ensures that deadlines are met, budgets are followed, and the final product is of the highest quality.

But it's not just about the people.

This is about the entire ecosystem. It's about the tools you use, the processes you follow, and the culture you cultivate. It's about building a playbook that everyone understands and can execute with precision. It's about regular training, continuous improvement, and always, always playing to win.

Moving from gut instinct to a data-driven approach is about recognizing the value of metrics and analytics in making informed decisions. It's about understanding that success in today's business environment requires a solid grasp of numbers, from Gross Margins and EBITDA to more granular data points that can provide insights into every aspect of your operations. The future of marketing agencies—and indeed all businesses—is grounded in data. It's what separates the winners from the losers. It's the difference between being a kid playing in the field and a professional team competing in the Super Bowl.

Your agency's journey from that playground scrimmage to the Super Bowl is what we're about to unpack in the coming chapters. We'll dive into the nitty-gritty of professionalizing each aspect of your operations. We'll explore how to refine your processes, how to build and nurture your team, how to leverage technology to your advantage, and how to create a culture of excellence.

So, are you ready to take your agency from the playground to the big league? Are you ready to embrace the discipline, the strategy, and the precision that comes with professionalization? Good. Because that's exactly what it takes to win in this game. Welcome to your journey to the NFL of marketing agencies.

Building a Data Driven Culture

Building a data-driven culture transforms how your marketing agency operates, infusing every decision with precision and insight. This shift may seem daunting, especially if you worry that adding structure and discipline will stifle the creative energy and fun that make your agency unique. However, integrating data-driven decision-making can actually enhance your workplace, making it more exciting and rewarding for everyone involved.

A data-driven culture starts at the top. Leadership must champion the value of data, demonstrating how it can drive success and improve everyone's performance. By setting clear expectations and showing the tangible benefits of data, you encourage your team to embrace this shift. Leaders should be transparent about the metrics they track, from key performance indicators (KPIs) to financial metrics like Gross Margins and EBITDA, and explain how these figures influence strategic decisions.

Data should not be seen as a constraint but as a powerful tool that complements creativity. Show your team how data can validate their ideas, identify trends, and predict campaign outcomes. For instance, using analytics to track the success of different marketing strategies can highlight which creative approaches resonate most with your audience, guiding future projects.

Emphasize that data-driven decisions lead to more effective and impactful creative work. When your team sees that their efforts are grounded in data, they gain confidence in their ideas and are more likely to take calculated risks, knowing there's a solid foundation backing them.

Introducing data-driven accountability may initially cause discomfort, especially for underperformers. However, this shift is crucial for fostering a high-performing culture. Top performers thrive in environments where their contributions are recognized and where everyone is held to the same high standards.

Creating a culture of accountability means setting clear, measurable goals for each team member, much like how NFL coaches set performance targets for their players. Just as a coach might set a target for a receiver to achieve a certain number of touchdowns, your agency can set specific goals for project managers, creatives, and strategists. These goals should be aligned with overall business objectives and supported by relevant data.

When team members know exactly what is expected of them and how their performance is measured, they are more focused and motivated. They understand how their work contributes to the agency's success and are driven to meet and exceed their targets. This clarity not only boosts individual performance but also fosters a sense of collective purpose.

A data-driven culture highlights the achievements of top performers, providing them with more opportunities to excel. High achievers often feel frustrated when their efforts are overshadowed by the inefficiencies or underperformance of others. By holding everyone accountable and recognizing excellence, you create an environment where top talent feels valued and motivated.

Data can also identify areas where top performers can further develop their skills or take on new challenges. Continuous performance tracking allows you to offer tailored development opportunities, ensuring that your best people continue to grow and contribute at a high level.

Integrating data into your culture doesn't mean sacrificing the fun that makes your agency a great place to work. In fact, a clear, data-driven plan for success can enhance the overall enjoyment of work. When team members see the direct impact of their efforts on the agency's success, they experience a greater sense of accomplishment and satisfaction.

You can maintain a vibrant, creative atmosphere by celebrating successes and using data to highlight wins. Regularly share insights and achievements, showing how data-driven decisions have led to successful outcomes. This not only reinforces the value of data but also keeps the team engaged and motivated.

Moreover, ensure that the process of collecting and analyzing data is as seamless and integrated as possible. Use user-friendly tools and platforms that make data accessible and actionable without overwhelming your team. By simplifying data processes, you allow your team to focus on their core strengths—creativity and innovation—while benefiting from the insights data provides.

Building a data-driven culture is essential for marketing agency owners who recognize that to reach their full potential, they must evolve much like I had to do. This transformation requires a fundamental shift in your mindset as the agency owner. By blending data-driven decision-making with your agency's creative spirit, you set the stage for sustainable growth and success, especially in these technologically changing times.

As you implement these changes, remember that professionalization and creativity are not mutually exclusive. A structured, data-driven approach provides the stability and clarity needed to foster innovation. Your team will thrive in an environment where their contributions are measured, recognized, and celebrated, driving both personal and collective success.

This book will guide you through the process of integrating data into every aspect of your agency, from strategy and operations to creative execution. By embracing this transformation, you ensure that your business not only survives but thrives in an increasingly competitive landscape. Welcome to the future of marketing agencies, where data-driven excellence meets creative brilliance.

The Path Forward

As we embark on this journey, you will witness the transformation of a marketing agency owner named Alex who, guided by a mentor, embraces the "Facts Not Feelings" approach. This story isn't just about business growth as it's as much about personal evolution and the profound impact of running a business based on data and numbers.

In the upcoming chapters, we'll dive deep into actionable lessons that will reshape how you manage your agency. You'll learn to harness the power of data, implement effective processes, and build a high-performing team. We'll explore how to leverage technology, set clear goals, and create a culture of accountability and excellence. Each chapter will provide practical insights and strategies that you can apply to your own business, ensuring that the journey you undertake alongside our protagonist is as enlightening and transformative for you as it is for them.

Our protagonist starts much like many agency owners—full of ambition and creativity but hampered by reliance on gut instinct and an aversion to the "boring" aspects of managing and running a business. This journey begins with the critical realization that gut instincts, while valuable, are not sufficient for sustained growth and success. The narrative follows Alex's struggles and breakthroughs as he shifts from making decisions based on intuition to a more structured, data-driven approach.

Early on, our protagonist meets a mentor who becomes a pivotal figure in his transformation. This mentor introduces him to the concepts of data-driven decision-making, showing how metrics like Gross Margins, EBITDA, and other key performance indicators (KPIs) are essential for understanding and steering the business. Through this guidance, our protagonist learns to view data not as a constraint but as a tool that enhances creativity and strategic planning.

As the story unfolds, you'll see the mentor's wisdom in action. He teaches Alex how to analyze performance data to uncover insights that drive better decisions. You'll learn alongside Alex how to set up data systems, interpret the numbers, and use these insights to refine strategies. This process demystifies the numbers, turning them from intimidating figures into vital tools for business growth.

One of the key transformations you'll witness is the shift in how the agency operates. You'll see how Alex is able to streamline operations, reduce inefficiencies, and improve overall performance. You'll see how adopting these processes helps in setting clear, measurable goals for each team member, much like how NFL coaches set performance targets for their players. This approach ensures that everyone knows what is expected of them and how their performance contributes to the agency's success.

Throughout the book, the importance of accountability is emphasized. The protagonist learns that by holding team members accountable to data-driven goals, they can identify underperformers and provide the support needed to improve or make tough decisions when necessary. This clarity and structure foster a high-performing culture where top talent can thrive and feel valued. High achievers often feel frustrated when their efforts are overshadowed by mediocrity. By holding everyone to a higher standard and recognizing excellence, you create an environment where top performers can shine.

Moreover, the story demonstrates that integrating data into the company culture doesn't mean sacrificing the fun and creative energy that make a marketing agency unique. On the contrary, a clear, data-driven plan for success can enhance the overall enjoyment of work. When team members see the direct impact of their efforts on the agency's success, they experience a greater sense of accomplishment and satisfaction. Celebrating successes and using data to highlight wins keeps the team engaged and motivated, maintaining a vibrant and dynamic workplace.

You'll also learn about the tools and technologies that facilitate data collection and analysis, making the process seamless and integrated. The protagonist's journey includes adopting user-friendly platforms that make data accessible and actionable without overwhelming the team. This allows everyone to focus on their core strengths—creativity and innovation—while benefiting from the insights data provides.

As you follow this story, you'll gain practical insights and actionable strategies to apply to your own business. You'll understand how to use data to drive creativity, improve efficiency, and achieve sustainable success. This book is for marketing agency owners who recognize that to reach their full potential, they must change, and the business must also change.

By the end of this book, you'll have a comprehensive understanding of what it takes to run a marketing agency like a professional. You'll be equipped with the tools and knowledge to make informed decisions, optimize your operations, and lead your team to new heights. This journey will not only transform your business but also lead to a more fulfilling and enjoyable professional life.

Are you ready to revolutionize your agency and unlock its full potential? Let's dive in and start this transformative journey together. Welcome to the path forward, where data-driven decision-making meets creative brilliance.

Chapter 1: The Wake-Up Call

Breaking Point: Speaking Hard Truths

Alex sat still, staring at nothing, the air thick with silence. He wasn't sure when he had last moved. The feeling of failure was now something tangible, heavy, like a thick fog pressing down on him. The silence in the house didn't help either. Every second felt louder, heavier. He wanted to scream, but instead, he just ran his hand through his hair, pulling at the strands slightly as if that would somehow keep the thoughts from running loose.

Emily watched him from the couch, her brow furrowed in worry. "Alex, please... sit down. You've been pacing for hours and you're going to wear a hole in the carpet if you don't calm the fuck down."

Alex didn't respond right away. His steps were quick, almost frantic. He couldn't sit, not with everything spinning out of control. "It's not just another client, Emily. This is everything. It's all falling apart, and I can't—" He cut himself off, a bitterness coating his voice.

Emily's fingers twisted the blanket in her lap. "What happened THIS time? I swear we keep having more and more of these conversations."

He stopped pacing for a moment and turned to her. "Same as always. Missed deadlines. Miscommunications. Projects going sideways. Clients leaving. What's new, right?" His laugh was hollow and empty.

"Alex..." Emily started, but she wasn't sure what to say. She'd seen him like this before, but tonight felt different. Darker. Deeper.

He dropped onto the edge of the couch, leaning forward, his hands gripping his hair. "It's me, Em. It's my fault. Every time we lose a client, every time something goes wrong... it's on me."

"You know that's not true babe," she said gently, but her words didn't seem to reach him.

He shook his head. "No. It is. I thought I could build this thing—this agency—into something big. Something that would eventually run itself. But every damn day, I'm there, micromanaging every single detail because nothing works without me. I'm... I'm failing."

His voice cracked and he let out a long, shaky breath, as if saying it out loud made the weight of it all even heavier.

Emily leaned forward, her eyes soft but filled with concern. "You're not failing. You've built something incredible, Alex. You just... you can't do everything yourself."

Alex looked at her, his eyes bloodshot, dark circles forming underneath. "But that's the thing. I have to do it all myself. Because if I don't, who will? No one takes responsibility. It's like they don't care."

She sat up, moving closer to him. "Maybe it's because they don't know how. Maybe you haven't given them the structure they need to succeed."

"Listen, I know you're doing your best. But I also know you really struggle with leading people. Sometimes you're too easy on them. Sometimes you're too soft on them. It's not that you're a bad person. I love you more than anyone on this planet. But I think it sounds like what you're doing is making your people feel just as lost and frustrated as you feel.", Emily finished.

Her words stung, not because they were wrong, but because they were too close to the truth. He had been running his agency on instinct, on gut reactions, never really stopping to build something solid.

"I've tried structure," he muttered, more to himself than to her. "I've tried systems. SOPs. None of it sticks. I'm constantly putting out fires, but no one else seems to see the flames until it's too late."

Emily sighed softly. "I hate seeing you like this, Alex. I really do. I wish I could do more..."

He looked at her, exhaustion evident in his eyes. "You can't fix this, Em. I don't even know if I can fix it."

Her heart sank hearing him say that. She had never seen him like this—defeated. "Maybe you just need help. Real help. Someone who knows how to turn this around."

"Help?" Alex echoed, his voice thick with skepticism. "You mean another consultant? Another person who doesn't understand my business but wants to charge me a fortune to give me generic advice? What I think I really need is a shrink who can prescribe me some strong antidepressants."

He said this half-joking, but at this point Alex would take any pill available to help him get from one day to the next.

"No," Emily said quickly. "I mean someone who's been where you are. Someone who knows what it's like to be in your shoes. A mentor, maybe."

Alex scoffed and leaned back, his head hitting the cushion with a thud. "A mentor? You think someone else has the answer to this mess? I look around at all my competitors and I truly feel like I'm the only one struggling right now."

She hesitated. "I don't know. But what I do know is that you can't keep doing this alone. I see how much it's tearing you apart. You've been carrying this weight for so long. Maybe it's time to let someone else share the burden."

He didn't respond right away. He felt the pressure building in his chest, that sinking feeling that everything was slipping out of his control. His pride screamed that he didn't need help—that asking for it would be admitting defeat. But the exhaustion in his bones, the ache in his heart, told him otherwise.

"I don't know, Em," he finally said, his voice barely audible. "I don't know if I can let go. I don't even know where to start."

Emily reached out, gently placing her hand on his arm. "You don't have to figure it all out tonight. But you need to give yourself permission to ask for help. You've been fighting this battle alone for too long. You're not Superman. You can't keep up this pace or I'm seriously worried you will have a nervous breakdown or worse."

He closed his eyes, the silence between them filled with unsaid words. He wanted to believe her, but the weight of failure was so heavy, it felt impossible to shake.

After a long pause, he nodded, but it wasn't a gesture of agreement. It was more a gesture of surrender— acknowledging that he couldn't keep doing this, even if he wasn't ready to face the solution just yet.

"I'll think about it," he mumbled, more for her sake than

his own. He stood up, pacing again, the nervous energy still too much to keep bottled up.

Emily didn't press him further. She knew pushing him would only make him retreat more. "We'll figure this out, Alex. You're not alone in this."

Her words were soft, reassuring, but they didn't touch the part of him that was spiraling. The part of him that couldn't shake the fear that he'd already lost control and that no one, not even a mentor, could help him regain it.

Revelations with Maya: Exposing the Cracks

Later that morning, Alex dragged himself into the office. The weight of his conversation with Emily clung to him like a heavy fog, clouding his thoughts. He couldn't shake the feeling that everything was crumbling. He felt it last night, and it hadn't disappeared. If anything, the daylight only made it worse.

He dropped his bag onto his desk and sat down, burying his head in his hands. He couldn't keep going like this. His eyes were gritty from lack of sleep, his chest tight with the suffocating pressure of his thoughts.

Maya knocked on the doorframe, watching him closely. "Alex?"

He looked up, eyes red-rimmed and exhausted.

"Did I catch you at a bad time?" Maya asked.

"I'm barely holding it together, Maya," he muttered. His voice sounded broken even to his own ears.

Maya stepped into the room, her expression one of deep concern. "What happened? You look like you didn't sleep at all."

Alex let out a bitter laugh. "Man I must look like shit, don't I. Well I didn't sleep if you must know. Couldn't stop thinking. I was just thinking about everything we have going on this week and... everything just... it's all falling apart."

Maya sat down across from him, listening.

"We lost another client," he continued, his voice shaking with frustration. "Same old story. Missed deadlines, mistakes, and now they're gone. But this time... it's not just about the client. I don't know how much longer I can keep this up."

Maya stayed silent, knowing he needed to get it all out.

"I told Emily I'm failing," Alex admitted, his words bitter. "I thought I was building something that could run without me, but I'm still here, micromanaging every damn thing because nothing works unless I do it myself. Every time we patch something, another part of the agency breaks."

Alex caught himself dumping his emotions on Maya. He knew that he shouldn't do that, but after working with her for the past 7 years, he felt she was the only one he could actually confide in.

Maya sighed softly. "You're not failing, Alex. This is what happens when we don't have proper systems in place. The team doesn't know how to succeed without guidance. We've been trying to wing it for too long."

Alex looked up, his eyes filled with defeat. "Emily said I need help. She thinks I should bring in a mentor. But I don't know if that'll change anything. Who's going to understand my business better than I do? How can anyone fix this mess?"

Maya's brows furrowed slightly. "Emily's right about one thing—you can't keep carrying all of this on your own. You've been doing everything, and that's not sustainable. But we need more than just another quick fix."

He rubbed his temples, the weight of it all pressing down harder. "What if I can't fix it? What if the problem is me?"

Maya shook her head. "You're not the problem, Alex. The way we're running things is. We've been reactive, dealing with problems as they come, instead of planning and setting up proper systems to stop them from happening in the first place. You need to step back, and we need to put processes in place that allow the agency to function without you having to solve every crisis."

"I've tried that," Alex snapped, his voice louder than he intended. "I've tried putting systems in place, and reading all the books, and hiring the gurus, but nothing sticks. Everything always falls back on me. No one follows through, and I'm left picking up the pieces."

Maya leaned in, her tone steady. "Because we haven't committed to real systems. We've been putting out fires instead of preventing them. You've been holding the whole thing together with duct tape, and it's not enough anymore. We need a solid foundation."

Alex stared at her, his frustration simmering just beneath the surface. He wanted to believe her. He wanted to believe that there was a solution, but after months—years—of trying to hold everything together, the idea of rebuilding felt impossible.

"I don't know if I can," he said quietly. "I don't know how to let go of this enough to trust someone else."

Maya's gaze didn't waver. "It's not about letting go, Alex. It's about giving yourself the tools to succeed. You're not losing control; you're building something that won't need you to be everywhere at once. You need someone who can help you see the bigger picture."

He rubbed his face, feeling the exhaustion wash over him again. "I told Emily I'd think about it. But honestly, Maya... I don't even know where to start. What if I can't pull this back together?"

Maya leaned back in her chair, her eyes never leaving his. "The fact that you're asking that question shows you're already on the right path. You don't have to have all the answers right now. But we need a strategy—a real plan. And that starts with admitting that doing it all on your own isn't working anymore."

Alex let out a long breath, leaning back in his chair. The room felt heavy, the weight of their conversation thick in the air. He didn't know if he could face the reality of everything Maya and Emily had been telling him, but deep down, he knew they were right.

"I'll think about it," he muttered again, his voice quiet and filled with uncertainty.

Maya didn't push. She knew he wasn't ready to commit to the idea of letting someone else in, but the seed had been planted. "Whenever you're ready, Alex. You don't have to do this alone."

Alex leaned forward in his chair, his hands gripping his knees as if the pressure of his own body weight might somehow ground him. His chest felt tight, the same unbearable weight he'd carried from home to the office clinging to him. Every breath felt shallow, like he was never quite getting enough air.

"I don't even know where to begin, Maya," he said, his voice low, almost a whisper. "Everything feels too big—too broken."

Maya nodded, her eyes scanning his face, taking in the exhaustion etched into every line. "It feels overwhelming because you're trying to hold it all at once. That's impossible. You need to stop carrying the weight of the whole agency on your shoulders."

Alex looked at her, his frustration peaking. "But how? Every time I step back, something goes wrong. Another client leaves, or the team screws up, or we miss another deadline. I can't keep cleaning up after everyone, but if I don't, we'll go under."

Maya took a deep breath, steadying herself before responding. "I understand, Alex. But the more you micromanage, the more dependent the team becomes on you. They're not stepping up because they don't have the chance. You're too busy solving every problem for them."

Her words cut deep, but Alex knew there was truth to them. He'd seen it happen again and again—he'd jump in to fix a mistake, only for another one to pop up somewhere else. It was like he was constantly playing whack-a-mole with the agency's problems, and it never stopped.

"It's funny because Emily said the same thing," he admitted, his voice barely audible.

Maya's expression softened. "She's right. You've been doing this for years—running yourself into the ground to keep everything afloat. But that's not sustainable, Alex. It's okay to ask for help."

Maya leaned forward, her tone gentle but firm. "You're not handing over the keys to the kingdom, Alex. You're bringing in someone who can help you see the bigger picture. Someone who can point out what's not working and give us the tools to fix it. You're not losing control— you're taking a step toward gaining control back."

He blinked, processing her words. For years, he had been so consumed with the idea of doing—of being the one to fix every crack in the system—that the thought of letting someone else take the reins, even for a moment, felt foreign. But there was no denying it anymore. His way wasn't working. It hadn't worked for a long time, and the cracks were getting bigger.

Alex rubbed his eyes, feeling the exhaustion pressing harder against his skull. "What if I bring someone in and they can't fix it? What if I let go and the whole thing falls apart anyway?"

Maya's gaze didn't waver. "What if they do fix it? What if letting go is the only way to actually save this agency, Alex? You're doing everything you can, but you can't keep burning the candle at both ends like this. You'll burn out, and then what? You'll have nothing left to give."

Her words hung in the air, heavy and unavoidable. He'd been afraid of this very thing—afraid that if he admitted he couldn't do it all, it would mean he'd failed. But hearing Maya say it so plainly... it wasn't failure. It was survival.

"I'm just tired, Maya," he admitted finally, his voice cracking under the weight of it. "I'm tired of fighting. I'm tired of feeling like I'm one mistake away from losing everything."

Maya responded. "I know you are. And that's exactly why you need to take a step back. You can't keep doing this alone. The agency is too big for one person to carry. If you don't bring in someone who can help us rebuild, we'll keep running in circles—until there's nothing left... nothing left of this company...and nothing left of you."

Alex swallowed hard, the lump in his throat making it difficult to speak. He hated the truth in her words. He hated how vulnerable he felt, how exposed. But Maya was right. Emily had been right. He couldn't keep carrying this alone.

"How or where do I even start?" he asked, his voice small, the question weighing on him like a boulder.

Maya smiled softly, the first sign of hope in the entire conversation. "You start by admitting you can't do it all. You start by letting go—just a little bit—and trusting that there are people out there who can help you. We'll find the right person, Alex. But we can't take the next step until you're ready to take your foot off the brake."

Alex closed his eyes, leaning back in his chair. The silence between them was heavy, but it was different now. It wasn't the suffocating weight of uncertainty, but rather the quiet aftermath of realization.

He wasn't sure if he was ready to let go, but for the first time, he was starting to understand that maybe... maybe it wasn't about letting go entirely. Maybe it was about loosening his grip just enough to let someone else help him hold on.

"I'll think about it," he muttered again, but this time, the words felt less like an excuse and more like a promise.

Maya nodded, her eyes soft with understanding. "That's all I'm asking."

Instincts vs Strategy: A Battle Within

Maya left his office and for the next few hours Alex pretended to work, but in reality he was replaying all his mistakes in his head. Every breath he took and every memory he recounted felt like he was sinking deeper, suffocating under the pressure that had been building for years.

This wasn't how it was supposed to feel, he thought, his hands trembling slightly as he ran them through his hair. Running the agency—his agency—should've felt like freedom. He should've been able to step back, to breathe, to enjoy the success that had taken years to build. But instead, every day felt like a new crisis, another fire to put out.

His eyes flickered across the room, landing on the hundreds of unread emails on his laptop—plus the reports, the complaints. Each one was a reminder that his once rock-solid instincts were now leading him astray. What used to be his greatest asset had become his greatest burden.

I can't keep doing this, he thought, his heart heavy with the realization. Instinct wasn't enough anymore.

Alex leaned forward, elbows on his knees, head in his hands. His mind raced, spiraling deeper into the thoughts he had been avoiding for weeks, maybe months. Where did I go wrong?

He had always trusted his gut. It was instinct that had carried him through the early days of the agency—making quick decisions, seizing opportunities, fixing problems on the fly. But now? His gut was leading him into disaster. Every instinctive decision seemed to lead to another mess, another missed opportunity, another client walking out the door.

Failure hung in the air like an echo. What if I'm the reason this is all falling apart?

He glanced at his phone on the desk. The temptation to reach for it, to call someone, was there—but his pride fought back. Asking for help felt like admitting defeat. How could he, the person who built this agency from the ground up, suddenly admit he couldn't handle it?

But the truth was staring him in the face. He couldn't handle it.

His chest tightened, and his thoughts turned back to Emily's words from the morning. You can't keep doing this alone, Alex. You need help. Her voice played in his mind like a song he couldn't escape. She had been right, and the more he thought about it, the harder it became to ignore the truth that was clawing at him from the inside.

I'm failing because I'm still trying to do this all by myself.

For years, he had told himself that trusting his instincts was enough. That if he just worked harder, pushed more, he could keep the agency afloat. But now, every move he made felt like he was fighting against quicksand—the harder he struggled, the deeper he sank.

By now the agency had been running for nearly 15 years and Alex had amassed an amazing team of 17 employees and numerous contractors and yet Alex knew that deep down he didn't trust anyone. The fear of failure still kept him awake at night and working all weekend, but the irony is that Alex knew that he was sabotaging his own success and yet he felt powerless to change his bad habits.

The desk in front of him blurred as his eyes grew glassy. He blinked, trying to push back the exhaustion that seemed to settle in deeper with every passing minute. He couldn't remember the last time he felt truly in control. The reports, the projects, the people—it all felt like it was crumbling, slipping away from him.

He stared at the stack of client files on his desk, his gut twisting with the familiar weight of failure. What good were instincts now? Instinct had been his compass when the agency was smaller, when he could touch every part of the business and fix problems as they came. But now? Now it felt like he was a captain trying to steer a ship through a storm with nothing but his bare hands.

Control was slipping away.

The pile of papers blurred in front of him again as the reality hit him: I've built this entire thing around me. My decisions. My instincts. And now, the very thing that had once made him a leader was the thing pulling him down.

I've made myself indispensable. The thought landed like a punch in the gut. By being everywhere, by making every decision, by solving every problem, he had created a situation where the agency couldn't function without him—and it was killing him.

His mind raced back to something Maya had said the other day: We need to stop putting out fires and start preventing them. Her words had been a casual suggestion at the time, but now they felt like a lifeline. Systems. Structure. Control that didn't rely on him being in every room, in every conversation.

But how? How does he shift from running everything on instinct to trusting something bigger, something he couldn't micromanage? What if systems failed? What if I let go, and it all falls apart? I'm the one who will suffer and I'll be the one who is remembered as a failure.

The fear gripped him, the idea of letting go sending a jolt of anxiety through his chest. But wasn't he already losing control? Wasn't everything already slipping through his fingers? His instincts had led him here—on the verge of collapse.

His phone buzzed on the desk, but he ignored it, his mind too deep in the spiral to pay attention. The thought of asking for help gnawed at him. The idea of reaching out to someone like Jordan—someone who had been where he was, who had built an even bigger agency—felt both like salvation and defeat.

What would Jordan think of me now? Alex's stomach twisted. He had always prided himself on being the one who figured things out on his own, the one who could navigate any storm. But now, it felt like the storm was too big, too relentless.

The ceiling above him felt like it was closing in, the fluorescent lights flickering slightly. He stared up at them, the cracks in the plaster mirroring the cracks he felt in his own mind, in the agency. Everything's falling apart, he thought, the words repeating in his mind.

He reached for his phone, his hand shaking slightly. Jordan's number was saved there, and it had been for years, but he had never thought he'd need to use it like this. For help. For guidance. For someone to tell him how to pull this thing back together.

I can't keep pretending like I have all the answers, Alex thought, his throat tightening. Because I don't.

Facing the Hidden Problem

Hours had passed, and Alex was still sitting at his desk, the light outside fading into a dull gray. The office was quiet, too quiet—just the faint hum of the air conditioning and the distant sounds of the city filtering through the windows. He hadn't moved since the last meeting. His team had left hours ago, but Alex remained there, stuck in a loop of thoughts.

Everything felt heavy—like an invisible weight pressing down on him from all sides. The papers on his desk, the emails he hadn't responded to, the client contracts waiting for approval. It all felt like too much.

He glanced at his phone, sitting face down on the desk. Jordan's number was still there, waiting. Just a call, he thought. Just one simple call. But the thought of dialing that number felt like the heaviest thing of all.

The idea of reaching out to Jordan made his stomach turn.. It wasn't just about the help—it was what it represented. Admitting failure. He had built this agency from the ground up, his hands in every part of it. Letting someone else in meant admitting he hadn't been able to manage it on his own. It felt like a betrayal of everything he had worked for.

He pushed the chair back, standing suddenly, the motion jarring in the stillness. Pacing across the office, his fingers drummed against the side of his leg. His mind spun in circles.

The truth was simple. Jordan knew what he was doing—he always had. When Alex had met him all those years ago, Jordan had been the calm in the storm, a man who had systems for everything. He remembered watching him navigate his businesses with a kind of ease that had always felt foreign to Alex, as if Jordan had unlocked some secret that Alex could never quite grasp.

But what if calling him now was too late?

The thought stabbed at him. What if he'd let things spiral so far out of control that there was no fixing it? What if Jordan looked at the mess Alex had created and turned him away? The fear stabbed at him, sharp and unrelenting.

His phone buzzed on the desk, and Alex walked back, picking it up almost absentmindedly. It wasn't Jordan—it was another client complaint. Another missed deadline. Another fire to put out. He tossed the phone back down, the frustration bubbling up inside of him again.

This couldn't continue.

The realization hit him like a kick in the balls. His gut had been driving the agency for too long—relying on instincts to solve problems, to push forward, but all it had done was bring him to the edge of collapse. His pride had kept him from seeing the truth for too long. It wasn't just about being good at what he did. It was about building something that could function without him at the center.

That's what Jordan had done.

All of Jordans' businesses ran like well-oiled machines, and here Alex was, barely holding it together with his ONE company. His jaw clenched as he imagined the conversation—the one he had been avoiding for months. What do I say? How do I explain this mess without looking like an idiot?

He could almost hear Jordan's voice, calm, composed: "It's not about the failures, Alex. It's about the systems that prevent them."

Alex shook his head. The knot in his stomach tightened. It wasn't that simple. Couldn't be.

He sat back down, the chair creaking under him as he leaned forward, head in his hands. What would he say when Jordan picked up the phone? Hey, man, I've screwed up everything. Can you fix it? No. That wasn't it. That wasn't the right way.

But something had to give. If he didn't call, the agency would keep spiraling, and he'd keep drowning in it. He had tried fixing things alone, and now everything was worse.

What was left to lose?

He stared at the phone again. The fear that had kept him from making the call was still there, but now, it felt different. Now, it felt like admitting the truth wasn't just about failure—it was about survival. Jordan had been where he was now many years ago. He knew what it felt like to be buried under the weight of an agency that wouldn't stop pulling at you from every direction.

The decision was already made, wasn't it?

He grabbed the phone, his hand steady now, and dialed. One ring. Two rings. His heart pounded in his chest. Then, Jordan's voice—calm, familiar—came through.

"Alex," Jordan said. "It's been a while....how the hell are ya? Crushing life as always?"

The tension in Alex's body shifted. His throat tightened, but this time it wasn't the weight of failure—it was relief, small and fragile. "Yeah," Alex said, his voice soft, almost breaking. "I can't say I'm crushing it. But I'm hanging in there. I've been meaning to call you for some time and I'm really sorry it's taken me so long to actually check-in with you."

Jordan's tone remained light, but there was something knowing beneath it. "Don't worry about it man. I know we're both busy these days. So what's going on? You sound like you've got a lot on your plate."

Alex paused, the words caught in his throat. This was it. The moment he'd been avoiding. He took a deep breath, forcing himself to speak. "Well, the truth is...I'm struggling, Jordan. Like really struggling. The agency... I thought I had everything under control, but I don't. It's falling apart, and I've been running it on instinct for too long. I don't have the systems, and I don't know where to start. I look at you and you have it all together and I... I just need help."

For what felt like an eternity, Alex explained all the mistakes he had made and all the problems he had created for himself. The worst part was Alex admitting, to someone he admired, that he was failing.

The mood of the conversation felt heavy now, but not as suffocating as before for Alex. Alex waited, heart racing, wondering how Jordan would respond.

Jordan's voice came through, steady and calm. "Alex, first off, you're not alone. And secondly, I am really glad you reached out to me. I've been exactly where you are today. It's not easy and for me the hardest part wasn't the business flailing it was me feeling like I had no one to go to. After I dug myself out of my own hole I promised to always be there if anyone needed help. So in a way, you're helping ME to fulfill on that promise to the universe."

Alex exhaled slowly. Relief. "You have no idea how much it means to me to hear you say that."

"We'll figure it out," Jordan said, his voice filled with confidence. "It's going to take time, but we'll get you back on track. You just have to be ready to commit to it, to the work it's going to take."

"I'm ready," Alex said, though part of him still questioned if that was true. But what choice did he have? "I'm ready," he repeated, this time with more certainty.

"Good," Jordan said, his tone lightening. "That's the first step. I'll come by next week, and we'll take it from there. One step at a time."

Alex hung up the phone, the weight in his chest lifting slightly. He wasn't alone in this anymore.

Chapter 2: Meeting the Mentor

First Encounter: A Lifeline Appears

Alex tugged at the collar of his jacket, his fingers brushing the frayed edge as he glanced up at the sleek café sign overhead. Rain had started to drizzle, dappling the sidewalk in front of him, and in that moment, it felt like the weather mirrored his internal state—gray, unsteady, uncertain.

He hadn't seen Jordan in years, and now here he was, about to ask for help from the man who always seemed to have life figured out. Jordan, the golden boy. Even back in college, Jordan had been that guy—the one who made it all look effortless, who never crumbled under pressure. And now? Now Jordan was living the life Alex had always wanted but never quite reached. Successful business, perfect family, and the kind of unwavering control over himself that Alex could only dream about.

Alex glanced down at his watch. Ten minutes early. He was always early, part habit, part nerves, but today the nervousness felt more like a lump in his throat he couldn't swallow down. Jordan had always been two steps ahead, and sitting down with him now—after all these years of silence—wasn't just about catching up. It was about admitting failure.

With a deep breath, Alex pushed open the door, the quiet chime of the bell punctuating his entry. The café was exactly the kind of place Jordan would pick—sleek, minimalist, everything in its place. Alex felt out of sync with it immediately, like his disheveled appearance didn't belong in the sharp edges and calm order of the space.

He scanned the room and spotted Jordan at the corner table, already seated, perfectly at ease. Jordan hadn't changed much. The same sharp features, the same relaxed posture, the same unshakable confidence radiating from him like a quiet forcefield. He looked like someone who'd never been late to anything in his life. Of course he was already here.

Alex's stomach twisted. How does he do it? How does Jordan manage to make life bend to his will while Alex was constantly putting out fires? Even the rain seemed to have avoided Jordan's perfectly tailored coat.

Jordan looked up, catching Alex's eye with a slight smile. It wasn't the kind of smile you gave an old friend—it was something more measured, as if Jordan had already assessed the situation, already knew why Alex was there. Always in control.

"Alex," Jordan said, as Alex extended his arm for a handshake, but Jordan pulled him in for a hug.

"It's been too long." Jordan replied and then looked Alex in the eyes like a proud father welcoming his son home from battle.

"Yeah," Alex replied, forcing a smile as they sat down. "It really has."

Jordan hadn't changed, and for a moment, sitting across from him, Alex felt small. Like the years that had passed had only widened the gap between them—Jordan with his pristine life, and Alex with his crumbling one.

Jordan leaned back in his chair, casual, but his eyes were sharp, scanning Alex like he was reading an open book. "So, how's everything going? The agency?"

Alex flinched at the question. He'd known it was coming, but the way Jordan asked it, so calm, so certain, made Alex feel like he was being exposed under a bright light. What could he say? Everything was falling apart? That he had barely slept in weeks, and his business was a constant avalanche of problems?

He swallowed. "Well, I mentioned a bit of it to you on the phone obviously...It's not great," he admitted, his voice quieter than he wanted. God, how did it come to this?

Jordan didn't seem surprised. He just nodded, the kind of slow, deliberate nod that made it clear he wasn't here for small talk. "Tell me what's going on."

Alex took a deep breath. "Honestly, it feels like everything's collapsing. I can't seem to get ahead of it. We're losing clients, deadlines are getting missed, and I'm just... I don't know. I thought I could handle it, but—" he hesitated, then sighed, letting the words hang in the air. "But I can't."

Jordan's expression didn't shift. He was unreadable, calm, but that was Jordan's way. He'd always had a way of making people feel like he knew what was coming next. "I get it," Jordan said simply.

Alex nodded and then shifted his eyes away in sham.

Jordan nodded again, leaning forward slightly, elbows resting on the table. "I've been there, Alex. It's easy to get stuck, to feel like you're just reacting to everything instead of controlling it. Especially when you're running on instinct."

Alex looked away again, staring down at the table and the glass of water in front of him that was beading with condensation. Every decision, every problem—Alex had handled it the way he always had: in the moment, on gut feeling. And now? It felt like his gut was leading him off a cliff.

"I built this agency from nothing," Alex said, his voice strained. "And now I feel like I'm watching it fall apart."

Jordan didn't flinch. "Because you're still trying to run it the same way you did when it was just you. You can't do that anymore. You're not working alone in your garage—

you've got a business now. Employees. Clients. You need more than instinct. You need facts."

Alex frowned. "Facts?"

"Facts, not feelings," Jordan said, his voice cool, but with an edge of authority. "You can't run a business this size based on how you feel in the moment. It has to be based on numbers, on data. Otherwise, you're just throwing darts in the dark."

Facts, not feelings. The phrase hung in the air, and Alex felt it sink deep into his chest. That's the problem, isn't it? He'd been running everything emotionally, reacting to problems instead of preventing them.

"You've always been good at thinking on your feet," Jordan continued, taking a slow sip of his coffee. "That's why you've gotten this far. But if you want to grow—if you want this agency to succeed—you need structure. You need systems. Something that can work without you making every decision."

Alex's throat tightened. Systems. Structure. That was the opposite of how he'd been running things. He was the fixer, the guy who stepped in when things went wrong. But maybe that was the problem. Maybe he'd built the entire agency around his ability to fix things, and now, as it grew, he couldn't keep up.

"I don't even know where to start," Alex admitted, the words barely above a whisper.

Jordan leaned back, his eyes steady. "That's why I'm here. We'll start with the basics—building a foundation that can handle growth. But first, you have to get out of your own way."

Alex looked up, feeling the weight of Jordan's words settle into his bones. For the first time in a long time, he saw the possibility of a way out, but it came with a heavy price—letting go.

Tough Love: The Brutal Diagnosis

Alex stared at the coffee cup in front of him, watching the swirl of cream and realizing, with a bitter pang, that it was the only thing in his life that felt remotely in control. His gaze shifted upward, meeting Jordan's eyes again. Jordan was sitting across from him, calm as always, his posture flawless, exuding that effortless confidence that had always made Alex feel two steps behind.

"Let me ask you something, Alex," Jordan's voice cut through the quiet hum of the café. His tone was casual, but Alex could feel the weight behind the question. "When was the last time you made a decision about your agency that wasn't based on how you felt in the moment?"

Alex shifted in his seat, feeling his defensiveness rise instinctively. "I—" He hesitated, his mind scrambling for an answer, but the truth was obvious. He'd been living in a constant state of reaction for so long, he couldn't even remember what it felt like to make a decision without that gut-wrenching anxiety gnawing at him. "I don't know," he admitted finally, his voice low.

Jordan didn't seem surprised. In fact, there was a hint of disappointment in his eyes, though he was too controlled to let it show fully. He simply nodded, leaning back in his chair, his fingers drumming lightly on the table. "That's what I thought."

Alex clenched his jaw, his frustration flaring up before he could tamp it down. "It's hard not to react when everything feels like it's falling apart around you," he snapped, the bitterness seeping into his voice before he could stop it. Jordan's calm was starting to grate on him— how could he sit there, so unaffected, while Alex's world was in shambles?

Jordan's expression remained unchanged, unflinching. He let the silence hang between them, as if waiting for Alex's outburst to fizzle out on its own. And it did. In a matter of seconds, Alex felt the weight of his frustration settle into exhaustion again.

"You're not leading your business, Alex," Jordan said, his tone deliberate. "Your business is leading you."

The words hit Alex harder than he expected, like a punch to the gut. He opened his mouth to argue, to defend himself, but nothing came out. He wanted to deny it, to say that wasn't true, that he was in control—but deep down, he knew Jordan was right. He wasn't leading anything. He was barely holding on.

Jordan leaned forward slightly, his eyes sharp, probing. "You've built your business on feelings—on instinct. And that works for a while, in the beginning, when everything's new and you're running on passion. But now? Now it's grown beyond what your instincts can handle. And that's why it's falling apart."

Alex felt his stomach churn. It was like Jordan was pulling the thoughts straight out of his head, giving voice to the fears he had tried so hard to suppress. Had he really built his agency on nothing but instinct? Had he let his emotions guide him for so long that it had become the very thing tearing everything apart?

"You're reacting to problems as they come," Jordan continued, his voice calm but unyielding. "You're putting out fires, making decisions based on fear, frustration, stress. And the moment those feelings shift, so does your direction. That's why you're stuck."

Alex swallowed hard, the tightness in his chest growing more pronounced. It was true—all of it. Every time he lost a client, every time a deadline was missed, he reacted. He panicked. He scrambled to fix it. And every decision he made in those moments was driven by the raw, pulsing anxiety that gripped him from the moment he woke up to the second he collapsed into bed.

"I get that," Alex muttered, trying to keep his voice steady, though he could feel the cracks forming. "But I don't know how to just—turn that off. I can't ignore how I feel, Jordan. These things matter to me."

Jordan's gaze softened, just slightly. "I'm not asking you to ignore your feelings, Alex. Your emotions matter. They help guide you in the right direction. But they can't be the foundation for how you run your business. Feelings aren't facts. And facts are what you need if you're going to turn this around."

Feelings aren't facts. The phrase lingered in Alex's mind, twisting around the edges of his thoughts. Feelings had been his compass for so long—how was he supposed to just set them aside? He'd built this agency on passion, on the fire that had driven him from the beginning. But now... now it felt like that fire had become an uncontrollable blaze, burning through everything he'd worked for.

Jordan reached into his jacket pocket, pulling out a small notebook. He flipped it open, scribbling something down before sliding it across the table toward Alex. "Here's what we're going to do," Jordan said, his voice matter-of-fact. "We're going to strip everything back. We're going to look at the numbers—the facts. What's working, what's not. And we're going to start building a foundation that doesn't rely on how you feel in the moment."

Alex stared at the notebook, the pages filled with neat, precise handwriting. It felt like a lifeline, but also like a challenge. Could he do this? Could he really admit that the way he'd been running his business all these years was wrong?

"What kind of numbers?" Alex asked, more to delay the inevitable than anything else.

Jordan smiled faintly, though there was no warmth behind it—just practicality. "Metrics, data, real numbers. Revenue, profit margins, client retention, employee productivity. Things that don't change just because you had a bad day. We're going to make decisions based on that, not on your feelings."

Metrics, data, numbers. Alex wasn't afraid of numbers, but he wasn't used to relying on them either. His gut had always been his guide—his instincts had always pulled him through when things got tough. But now? Now it seemed like his gut had led him off a cliff, and he was just waiting for the fall.

"You're not just reacting to the problems in front of you anymore," Jordan continued, his tone still calm but firm. "You're going to look at the bigger picture. You're going to see where the real issues are—the bottlenecks, the inefficiencies. And once you see the numbers, the feelings lose their power."

Alex's hands tightened around the edges of the notebook. Lose their power. Could it really be that simple? Could the constant, gnawing fear and anxiety he felt every day really lose its grip on him if he just focused on the facts?

He glanced up at Jordan, who was watching him closely, waiting for the realization to fully sink in. Jordan had always been like this—calm, composed, and in control. He'd never let his emotions get the better of him. And now, here he was, offering Alex a way out of the chaos.

But it wasn't just about the numbers, was it? It was about Alex. It was about him admitting that the way he'd been running things was wrong. That he'd been relying on feelings because it was easier than facing the hard truth.

"I don't know if I can do this," Alex said quietly, his voice barely above a whisper. "I've been running this business on instinct for so long... I don't know if I can just flip a switch and change everything."

Jordan's eyes softened, though his tone remained firm. "You're not flipping a switch, Alex. This isn't going to happen overnight. It's going to take time, and it's going to take discipline. But if you want this business to survive— if you want it to grow—you have to stop reacting and start leading."

Alex felt the weight of Jordan's words settle deep into his chest. Reacting versus leading. For years, he'd been reacting to every problem that came his way, thinking he could fix it with sheer force of will. But now... now it was clear that he wasn't leading. He was surviving.

"I haven't been a leader," Alex admitted, his voice hollow. "I've just been... getting by."

Jordan nodded, his expression sympathetic but unyielding. "And that's okay, Alex. But now you have the chance to change that. You can start building something real—something sustainable. But you have to be willing to let go of the way you've been doing things."

Alex stared down at the notebook again, the weight of the decision pressing on him like a boulder. He wasn't just facing a shift in how he ran his business—he was facing a shift in who he was. For years, he'd relied on his gut, on his passion, on the belief that if he just worked harder, he could make it work. But now, that belief had led him here—on the brink of collapse.

"I don't even know where to start," Alex admitted, his voice cracking under the pressure of it all.

Jordan didn't waver. "You start with the numbers. You start with the facts. And you let the facts guide you.

Alex stared down at the notebook Jordan had slid across the table, his fingers hovering above the cover as though the small action of picking it up might somehow commit him to everything it represented. The pages were filled with promises of control, of numbers and metrics, and clear, calculated decisions—things that had never been a part of how he ran his business. Could it really be that simple? Could trading his gut for cold, hard facts turn everything around?

Jordan's voice broke through his thoughts, steady and calm as ever. "It's not about ignoring your instincts, Alex. It's about understanding that your instincts have their place—but they can't drive the bus anymore. They're there to guide you, sure, but you need to be steering with facts."

Instincts. Alex had relied on them for so long that the idea of stepping away from them felt like giving up part of himself. He had built this agency from nothing, trusting his gut, putting in the hours, making it work through sheer force of will. But now... that same instinct had driven him into chaos.

Alex rubbed his temples. "I've tried systems before, Jordan. I've tried putting processes in place, and every time, it just... falls apart. I end up back where I started, fixing everyone's mistakes, handling all the crises."

Jordan watched him closely, his gaze sharp but not unkind. "And why do you think that is?"

Alex frowned, the question hanging in the air between them. Why did every attempt to build structure in his agency fail? Why did everything always fall apart the moment he tried to step back? Was it the team? The clients? Or was it... him?

"I don't know," Alex said finally, the words heavy with defeat. "Maybe I'm just not cut out for this."

Jordan's expression didn't change, but there was a flicker of something in his eyes—something Alex couldn't quite place. "Listen man, you need to cut that self-pity crap out. I know you. I know how great you are. It's not that you're not cut out for it, Alex. It's that you've been running everything yourself for so long, you haven't given your team the chance to succeed without you. You've built your business on a foundation of reaction, not structure. That's why everything crumbles the moment you try to step back."

A foundation of reaction. That's exactly what it felt like. He had been scrambling for so long, trying to keep everything together, reacting to every problem as it came, that he had never actually built anything solid.

Jordan leaned forward, his voice lowering slightly, his tone more intense. "Let me be clear, Alex. If you keep running your business like this, it's going to break. And not just in small ways—it's going to collapse. You're going to burn out, your team is going to burn out, and before you know it, you won't have an agency left to save, and you might not even have a family either. I've seen it happen to other people many times."

The weight of Jordan's words settled heavily in Alex's chest, the tightness growing more suffocating by the second. Collapse. Burnout. Losing Emily. He could feel it, creeping closer every day, the edges of his world fraying more with each passing crisis. How long had he been holding everything together with nothing but duct tape and determination?

"I don't know how to fix it," Alex admitted, his voice cracking slightly. The vulnerability in his words made him feel exposed, raw, but there was no point in pretending anymore.

Jordan nodded slowly, his eyes never leaving Alex's. "I know you are. And that's exactly why we need to change things now. Not tomorrow, not next week—now. You're not just tired, Alex. You're running on empty. And if you don't put systems in place to help you manage this business, it's going to drain you until there's nothing left."

Alex swallowed hard, the tightness in his throat making it difficult to breathe. Jordan's words weren't just a warning—they were a lifeline. He could feel it, the way they cut through the haze of doubt and fear, offering him a way out of the chaos. But could he take it? Could he really step away from the way he'd been doing things for so long and trust that the numbers, the facts, would guide him in the right direction?

"I don't know if I can let go," Alex said, his voice barely above a whisper. The admission felt like a failure in itself—letting go meant losing control, and control was the only thing that had kept him afloat for this long.

Jordan leaned back in his chair, his gaze softening just slightly. "Letting go doesn't mean giving up control. It means building something that doesn't rely on you being there for every decision, every fire. It means trusting the systems to work so that you can focus on leading—not reacting."

Alex closed his eyes, the weight of the conversation pressing down on him like a physical force. Trusting the systems. Trusting the numbers. It felt like such a foreign concept, like stepping into a world he didn't quite understand. But maybe that was the point. Maybe that was the first step toward turning things around.

Jordan's voice broke through the silence again, steady and calm. "I'm not asking you to change overnight, Alex. This is going to take time. It's going to take work. But if you're serious about fixing this—if you're serious about building something sustainable—you have to start trusting the facts."

Alex nodded slowly, his hands still gripping the edges of the notebook. "Okay," he said finally, the word feeling both fragile and full of possibility. "Okay. Let's do it."

Jordan smiled, though there was a hint of something more behind it—a challenge, a test. "This is just the beginning," he said, his voice firm. "You're going to hate this process. There will be days when you want to quit, when you feel like it's too much. But if you want real change, Alex—if you want to build something that lasts—this is the only way."

The words hung in the air between them, heavy with the weight of everything Alex had been avoiding for so long. He could feel the shift happening, slowly but surely, as though the ground beneath him was finally settling after years of instability.

For the first time in a long time, Alex felt something stir inside him—not panic, not fear, but resolve. He wasn't sure if he was ready for everything that lay ahead, but for the first time, he could see the path forward. It wasn't clear, and it wasn't easy, but it was there. And that was enough.

"I'm ready," Alex said again, though this time, the words felt a little more certain.

Jordan nodded, his expression serious. "Good. Because now the real work begins."

Alex sat there, staring at the spreadsheet in front of him, his instincts telling him to run away. Numbers. He'd always hated them—hated how they made him feel small, boxed in, like they were choking the life out of everything he loved about his business. He wasn't a numbers guy. He was a creative, a visionary. That's what had built his agency, not some sterile figures on a screen.

But here he was, sitting across from Jordan, who was flipping through the documents like it was a normal Tuesday, his eyes scanning the data with laser focus. Alex tried to follow along, but his mind was wandering, jumping from one thought to the next, barely able to focus on the task at hand. This is why I avoid this.

Jordan's voice cut through his thoughts. "Alex, your numbers are screaming at you. You're just not listening."

Alex winced, feeling the weight of that statement. He knew Jordan was right, but numbers had never spoken to him the way they seemed to for Jordan. He wasn't wired like that. Jordan could glance at a balance sheet and immediately know what was wrong. Alex? He saw chaos. Rows of figures that meant nothing to him.

"I've been avoiding this," Alex admitted quietly, his hands clenched into fists on the table. "I know the numbers are bad, but every time I look at them, it feels... overwhelming. Like I'm drowning in data, and I don't even know where to start."

Jordan didn't respond immediately. He leaned back in his chair, his gaze steady but thoughtful. "You've been avoiding it because you haven't yet learn how running the business from the numbers actually gives you MORE freedom to do what you love. Don't get me wrong. I get it that shooting from the hip and using your instincts to make decisions is exciting. And for a while, that worked. But Alex, you can't run a business this size on gut feelings alone. Not anymore."

Alex felt the familiar tightness in his chest—the panic that always rose when someone talked about numbers, about structure. Why couldn't it just be simple? Why couldn't he keep running things the way he had in the beginning, when everything felt exciting and manageable?

Jordan slid the laptop back toward him, his expression softening slightly. "I get it. You're a visionary, not a numbers guy. But here's the thing, Alex—being a visionary doesn't mean ignoring the facts. It means using them to fuel your vision. You can't lead your team if you don't know where you stand."

Alex let out a long breath, his frustration building. "I've tried, Jordan. I've tried looking at the numbers, but every time I do, it's like they don't make sense. I might as well be looking at ancient hieroglyphs for that matter. I don't even know what I'm supposed to be looking for 90% of the time."

Jordan leaned forward, his tone firm but not unkind. "That's because you're not looking at the right numbers. You're trying to see everything at once, and it's overwhelming you. We're going to strip it back to the basics. Focus on what matters most. Trust me, this is much easier than you think it is."

Alex put his hands over his face, his mind still racing, then asked, "Okay... so what matters most?"

Jordan pulled out a notebook and started writing. "First, your Gross Margin. You're at 40% right now, and that's a problem. A healthy Gross Margin should be around 70%. Your Cost of Goods Sold is way too high—60% is bleeding you dry. Cost of Goods Sold is basically how much you're spending to get the client work done. That's the first thing we're going to tackle."

Alex nodded, feeling a knot form in his stomach. Bleeding me dry. The words hung heavy in the air. He had known things weren't great, but hearing it laid out so plainly made him feel like he had failed at something fundamental. He was supposed to be running a thriving business, and instead, it was barely hanging on.

"And then there's your overhead," Jordan continued. "You're spending around $90,000 a month, which wouldn't be a problem if your margins were better. But right now, after all expenses are paid, you're spending almost as much as you're bringing in. You're surviving on razor-thin margins, and that's a dangerous place to be."

Alex's heart sank further. Surviving on razor-thin margins. That wasn't what he had set out to do. He had built this agency to give himself freedom, to create something that could thrive without him constantly propping it up. But now? Now it felt like the business was an anchor, dragging him down, making it harder to breathe every day.

Jordan's voice softened, but the weight of his words didn't. "Alex, you've been running your business on feelings and I get it, that's what got you here. But it's not going to get you where you need to go. You need to start making decisions based on facts, not feelings. That's the only way to turn this around."

Facts, not feelings. The phrase that Jordan had mentioned several times before and it stuck in Alex's mind, repeating over and over again like a mantra. Facts, not feelings. He had built his entire business on feelings—on passion, on instinct, on gut decisions. And now Jordan was telling him that was the very thing that was tearing it apart.

Jordan pulled out another sheet of paper, this one filled with neat columns of data. "Let's look at your client retention next. Your churn rate is around 10% per month, which means you're losing one out of every ten clients each month. That's a huge problem. For your industry it should be about 3 out of every 100 clients so you're more than 3x over the best practice. Then you're trying to outsell this churn rate and it's really hard to keep that up."

Alex winced. He had known the churn rate wasn't great, but he hadn't realized it was that bad. "Clients leave for all sorts of reasons," Alex muttered, trying to defend himself. "Sometimes it's budget cuts, or they find a cheaper option, or they just... move on."

Jordan's gaze didn't waver. "And sometimes they leave because they're not getting what they need from your agency. That's what we need to figure out. Why are they leaving? What's the pattern?"

Alex rubbed his face, feeling the exhaustion wash over him again. Patterns. Numbers. Facts. It all felt like a foreign language, one he wasn't sure he'd ever be fluent in. But there was no denying it anymore—he couldn't keep running from this. He couldn't keep relying on instinct and hoping for the best.

"We're going to figure it out," Jordan said, his voice steady. "But it starts with looking at the facts. No more guessing, no more reacting. We're going to dig into the data and find out where the problems are."

Alex nodded slowly, the weight of the conversation still pressing down on him. "Okay. So where do we start?"

Jordan smiled faintly, though it didn't reach his eyes. "We start by accepting that the way you've been doing things isn't working. That's the hardest part, Alex—admitting that your instincts have been leading you astray. But once you do that, you can start building something real."

Alex felt the sting of Jordan's words, but he also felt something else—something he hadn't felt in a long time. Hope. It was faint, barely there, but it was enough to keep him sitting at the table, to keep him listening, to keep him from walking away.

"Alright," Alex said, his voice quieter than he intended. "I'm ready. What's next?"

Jordan leaned forward, his gaze intense. "Next, we get into the numbers. We fix your Gross Margin, your COGS, your overhead, your churn. We figure out why clients are leaving and how to keep them. And we build a structure that doesn't rely on you putting out every fire."

Alex nodded again, feeling the knot in his stomach tighten. Structure. Systems. These were the things he had always resisted, the things that felt like they would take the soul out of his business. But now? Now they felt like the only things that could save it.

Jordan stood up, pulling out a few more documents. "We're going to take this one step at a time. It's not going to be easy, and there are going to be days when you want to quit. But if you're serious about saving this business, Alex—if you're serious about turning things around—this is the way."

Alex swallowed hard, feeling the weight of the decision pressing down on him. But for the first time, he felt like he had a path forward. It wasn't going to be easy, but at least it was a path.

"I'm ready," Alex said again, though this time, the words felt a little more solid.

Jordan gave him a small nod. "Good. Then let's start."

Alex exhaled, his fingers still twitching slightly. This is it, he thought, no more excuses, no more running. For years, he had led with instinct, letting emotions pull him in every direction. Now, Jordan was giving him a lifeline, a structure that didn't just depend on hunches or gut feelings.

Alex couldn't help but feel a wave of discomfort as Jordan dove back into the numbers. Gross Margin, COGS, churn rate—all concepts that had always felt secondary to the creative drive that had fueled him. But now, they were unavoidable. The room seemed to close in on him as Jordan laid out the path forward. Each number Jordan revealed felt like a blow to the gut.

"You have the vision, Alex," Jordan said, his tone still calm but insistent. "But vision without structure is chaos. It's like building a house on sand—it might look good for a while, but the foundation won't hold up when the pressure hits. And right now? You're sinking."

Alex leaned back in his chair, the weight of Jordan's words pressing down harder. Sinking. That's exactly what it felt like. The agency that once filled him with excitement now felt like a ship taking on water. How did it get this bad?

"I hear you," Alex muttered, though his mind was still racing. "But how do I turn this around? I don't even know where to start."

Jordan leaned forward. "First, we're going to look at every service you provide and reprice it. There's a good chance you're undercharging, especially given the quality of work your team is putting in. Then we'll evaluate your vendors, your contractors, and every cost associated with delivering services and see where we can reduce those expenses. And finally we need to look at your SOPs..."

"Or lack thereof." Alex smirked

"Exactly, and we will find the inefficiencies in how we are doing our fulfillment and figure out how to be more efficient and more profitable...without compromising on the quality. We MUST continue to get our clients' results." Jordan concluded.

Alex stared at him, feeling overwhelmed already. Repricing? Cutting costs? It sounded like a monumental task, and Alex had never been one to focus on operational details like that. But Jordan made it sound so simple, like it was just another part of the puzzle.

"And while we're doing that," Jordan continued, "we're going to look at the client retention strategy. Losing 10% of your clients every month is not sustainable. If we don't address that, you're stuck in a cycle of constantly trying to replace lost revenue. Most of the time when your clients cancel it's a) we aren't getting them the leads they expected and/or b) they aren't feeling the "love" from us. Essentially, they don't feel like we care about them or their business. So we need to figure out why clients are leaving and fix those issues."

Alex felt a twinge of guilt. He had always chalked up client churn to external factors—budget cuts, clients outgrowing the agency, or competitors undercutting them. But Jordan's words cut deeper. Maybe it wasn't just the clients. Maybe it was the agency. Maybe it was him.

"I always thought churn was just part of the business," Alex said, his voice softer now. "Clients come and go, right? I can't control everything."

Jordan's gaze didn't waver. "You can't control everything, no. But you can control how your clients experience your agency. If they're leaving, it means there's a gap between what they expect and what you're delivering. And that gap? That's on you. It's your job to close it."

"So, how do we close it?" Alex asked, his voice more desperate than he intended.

Jordan gave a small nod, as if he had been waiting for that question. "We start with communication. Regular check-ins, proactive updates. Your clients need to feel like you're ahead of the curve, that you're anticipating their needs, not just reacting to problems. And we need to make sure your team is aligned with that vision. Everyone, from account managers to project leads, needs to be in sync. No more silos."

Silos. That word had come up a lot recently. His team was talented, but they often operated in isolation, and Alex spent more time trying to stitch everything together than actually leading the agency.

"Okay, I get that," Alex said. "But what about the team? How do I get them on board with this? They're already overloaded."

Jordan's expression softened. "We're going to take a close look at your team, Alex. Who's performing, who's not, and what roles are overlapping. You might need to make some hard decisions. Some people might not be right for the team anymore, and others might need to step up into bigger roles."

Alex felt a cold sweat break out. Hard decisions. He knew what that meant. There were people on his team—people he had hired, people he had worked alongside for years—who weren't pulling their weight. He had avoided confronting it for too long, always finding excuses or giving them the benefit of the doubt. But if this business was going to survive, he couldn't afford to ignore the problems anymore.

Jordan tapped the table with his pen, drawing Alex's attention back. "We're not just looking at numbers here. We're looking at everything—your people, your processes, your clients. This isn't about tweaking a few things. This is about rebuilding the foundation."

Alex nodded slowly, feeling the weight of Jordan's words sink in. Rebuilding the foundation. That's what this was going to take. He couldn't keep putting out fires. He couldn't keep hoping things would magically improve on their own. He had to step up. He had to lead.

"Alright," Alex said, though his voice felt shaky. "I'm in. What's next?"

Jordan smiled faintly, though there was no sense of relief in it. "Next, we start the hard work. We'll dive into the numbers deeper, start evaluating your team, and build a roadmap for getting your Gross Margin to 70%. It's going to take time, Alex. But if you commit to this, we can turn things around."

Alex exhaled, feeling the weight of the decision settle on him. Commit. That was the word Jordan had used. And for the first time in a long time, Alex felt like maybe—just maybe—there was a way out of this mess. It wouldn't be easy. It wouldn't be quick. But there was a path forward.

Jordan stood up, gathering his papers. "We'll start tomorrow," he said. "I'll send over a breakdown of what we need to tackle first. In the meantime, think about your team. Think about who's ready to step up and who might need to go. These are the decisions that will shape the future of your agency."

Alex nodded, the knot in his stomach tightening again. Who needs to go. The words hung in the air like a storm cloud. He had always prided himself on being loyal to his team, on building a culture of trust and camaraderie. But now, he could see that loyalty might be what was holding him back. If the agency was going to survive, he couldn't afford to carry dead weight.

Jordan turned to leave, but then paused at the door. "Alex," he said, his voice softer now. "This is going to be hard. Harder than anything you've done before. But if you stick with it, you'll come out stronger on the other side. You'll have a business that runs itself, a team that's aligned, and the freedom you've been chasing for years. But you have to be willing to let go of the way things have always been."

Alex nodded again, though the words caught in his throat. "I will."

Jordan gave him one last look before walking out the door, leaving Alex alone in the office. The silence settled in, but it wasn't the suffocating silence of failure that had been haunting him for months. It was different. It was the silence before something new, something harder, but also something better.

Alex stared at the numbers on the screen, his heart pounding in his chest. This is it, he thought. The moment everything changes. He wasn't sure if he was ready. But for the first time, he wasn't sure he had a choice.

A Radical Approach: Rethinking Success

The next week, Jordan and Alex met in the office. It was the first time that Jordan had come to the office since their grand opening several years earlier. The two of them sat in the conference room and even though it was Alex's office, he felt like an imposter sitting next to Jordan who knew the inner workings of the agency and just how close they were to losing it all.

Jordan, always prepared and in control, handed Alex a binder with a small stack of checklists. Alex sat staring at the checklists Jordan had placed in front of him. Each box seemed to mock him. Structure. Systems. Numbers. These were the things he had spent his career avoiding, and now, they were staring him down like a brick wall he had no choice but to climb.

Jordan leaned back in his chair, watching Alex carefully. "I know what you're thinking," he said, his voice calm but firm. "How can these help ME? My business is different and that's why I haven't been able to find anything that actually works in MY business. But trust me when I say that these checklists will change your life once you see all the gaps you've been overlooking."

Alex sighed, running a hand through his hair. "It's just... this isn't how I thought it would go. I built this agency with my own hands and everything feels so customized. I know it's what got me into this problem in the first place, but I guess I am just a little skeptical."

Jordan leaned forward, placing his elbows on the table. "That's exactly the point, Alex. You've built something incredible, but it's not sustainable if it all relies on you. You're the visionary, and that's great, but you're too deep in the weeds. We need to get you back to the big picture."

Alex swallowed hard. Big picture. That's what he had always prided himself on, but lately, he couldn't see past the fires he was constantly putting out. His mind flashed back to the client calls, the missed deadlines, the sleepless nights.

"I'm looking at all these items on your checklist and honestly don't even know where to start," Alex admitted, his voice barely above a whisper. "There are so many things that I am missing and I'm not like you, Jordan. You've got your shit together. Everything in your life is... perfect."

Jordan's eyes softened, and for the first time, Alex saw a flicker of empathy. "My life isn't perfect, Alex. I've just built systems that work. That's what keeps things running smoothly, whether it's in my business or at home. But trust me, I've been where you are."

Alex shook his head. "I don't know if I can do this. The numbers, the checklists... it all feels like—"

"—like the opposite of who you are," Jordan finished for him, his tone understanding. "But that's why you need it. You've been running this agency by feel for so long that the cracks are starting to show. And that's okay. But now it's time to shift gears."

Jordan flipped through the binder again, landing on a page that showed a series of charts—gross margin, cost of goods sold, overhead, and profit margins. "Let's talk numbers."

Alex's stomach tightened. Numbers. They had always been his weak spot. Every time he tried to sit down and work through the financials, his brain would shut down, his focus scattering like leaves in the wind. It was one of the reasons why he had avoided looking too deeply at the finances—he didn't want to face the reality of how bad things might be.

"Last week we talked about how your gross margin is at 40%," Jordan said bluntly, his finger tapping the chart. "And now you know that that's way too low. You should be at 70%. Now you know that this is one of the biggest culprits that is killing your profitability."

Alex started to feel a headache starting to form. "I know it's bad. But every time I try to sit down and figure it out, it's like my brain just freezes. I get overwhelmed."

Jordan nodded, as if he had heard this many times before. "That's your ADHD talking. You're a creative, a visionary— you thrive on new ideas and big-picture thinking. But when it comes to details like this? Your brain wants to shut down. That's why you need the systems. They do the heavy lifting for you."

Alex let out a long breath. He knew Jordan was right, but the idea of diving into these numbers—numbers that represented his failures—felt like wading through quicksand. "I've been avoiding this for so long. What if it's worse than we think?"

Jordan shrugged. "It probably is actually because you haven't been tracking your expenses properly. But the good news? You're not alone. We're going to tackle this together, one step at a time. And once we get the numbers right, everything else will fall into place."

Alex nodded slowly, feeling the weight of the task ahead. His mind flashed to his team—Maya, who had been with him through thick and thin, and the rest of his employees, who were all struggling under the weight of a poorly run agency. He had let them down by not stepping up as a leader. He had let his own fear of failure and numbers keep him from giving them the structure they needed to succeed.

"Okay," Alex said finally, his voice shaky but determined. "Where do we go next?"

Jordan smiled slightly, flipping to the next page in the binder. "We start by breaking it down. First, we get your gross margin to 70%. As I mentioned last week, we're going to have to raise prices, cut unnecessary costs, and find efficiencies. Then we tackle the team—defining roles, holding people accountable, and getting rid of dead weight."

Alex winced at the thought of firing people. "I hate the idea of letting people go."

Jordan's expression softened. "I get it. But sometimes, that's the only way forward. You can't keep carrying people who aren't pulling their weight. It's dragging everyone down, including you. But here's the thing. We will even make decisions about people using our "Facts Not Feelings" method so while it will still suck to let people go we can at least be confident that those who didn't make the cut were evaluated fairly."

Alex nodded, though the knot in his stomach tightened. He knew there were people on his team who weren't performing, but he had always avoided confronting them.

He didn't want to be the bad guy. But now, he realized that avoiding the issue had only made things worse.

"I'm going to need Maya's help with this," Alex said quietly. "She's always been the one keeping things together."

Jordan nodded. "Maya's a key player in this. She'll help you implement the systems, but you need to be the one driving the change. This can't just be her fixing your problems—it has to come from you."

Alex swallowed hard. It has to come from you. The words echoed in his mind. For years, he had been reactive, putting out fires and hoping things would get better. But now, he had to be proactive. He had to lead.

Emily happened to be in the office that morning and walked into the conference room at that moment, her presence instantly grounding Alex.

She smiled brightly as she realized surprisingly that she had interrupted Jordan and Alex's meeting and, placing a gentle hand on Alex's shoulder, said "It seems like I interrupted to business titans planning to take over the world. How are the Death Star plans coming along?"

Jordan turned to her and welcomed her with a huge hug.

"Emily it is so nice to see you!" he joyfully expressed. "I really appreciate you letting me steal your husband for a couple hours".

Alex looked up at her, his chest tightening with a mix of anxiety and relief. "Hey babe. We're just diving into a bunch of papers. It's... a lot. But Jordan's helping me get through it."

Emily nodded, her eyes filled with understanding. "Well I know you're in good hands and that he'll help you to get through this."

Alex felt a lump form in his throat. He had been so wrapped up in his own stress and failure that he had forgotten he wasn't in this fight alone. He had people around him— Emily, Maya, Jordan—who believed in him, even when he didn't believe in himself.

"Ok, well you two guys behave yourselves. I just had to grab my credit card from your wallet, Alex. You took it to buy groceries the other day and never gave it back, you dick."

"Oh man. I'm sorry about that. Totally slipped my mind" Jordan replied.

"Don't worry about it. I know you're going through a lot right now." And as she said those words she gave a soft, loving glance to Alex and walked out of the conference room.

Jordan cleared his throat, drawing their attention back as she left. "Alright, let's get back to it. We're going to create a timeline for implementing these changes. One step at a time. First, we fix the numbers. Then we address the team. After that, we can start focusing on growth."

Alex nodded, already scribbling notes. "I'll start reviewing the pricing and contracts. We can identify where we're losing money and where we can raise prices without losing clients. Luckily we have time tracking in place so I think I can have Maya start calculating those numbers. Not sure how she will do it but she can figure it out."

"Good," Jordan said. "Alex, then you and Maya will get clear on your team. Who's performing, who's not, and who needs to be let go. No more avoiding those conversations."

Alex took a deep breath, feeling the weight of responsibility settle on his shoulders. He wasn't just fixing the agency for himself—he was doing it for his team, for Emily, for everyone who had believed in him. And for the first time in a long time, he felt a flicker of hope.

"Let's do it," Alex said, his voice steadier now. "No more avoiding. No more running. We're going to make this happen."

Blueprints and Checklists: Building the Foundation

Jordan introduces Alex to the checklist system, explaining how structured growth, driven by data and metrics, leads to sustainable success. Alex starts to see how his gut-driven approach has limited his agency's growth. Jordan's no-nonsense approach leaves Alex reflecting on how deeply he must change his mindset.

The office was quiet after Jordan left, the weight of the day's conversations pressing down on Alex like a suffocating blanket. The checklist, the numbers, the stark realities— they all buzzed in his mind, each one demanding his attention, each one a reminder of how far he had to go. He felt like he was staring at the wreckage of something he had built with his hands, now shattered, and the daunting task of picking up the pieces felt overwhelming.

Alex walked back to his office and sat at his desk staring at the binder that Jordan had given him. The rows of metrics mocked him: 40% gross margin, 60% cost of goods sold, each a glaring red flag. For so long, these numbers had been just another set of data points he ignored, hiding behind the comfort of creativity and instinct. Now, they were the smoking gun, evidence that the very foundation of his business was crumbling.

The door creaked open, and Maya poked her head in, her eyes scanning the mess of papers strewn across Alex's desk. "Hey," she said softly, "You alright?"

Alex sighed and gestured for her to come in. "I'm not sure, honestly. Jordan left me with a lot to think about. And a lot for you and I to do. It feels like everything I thought I knew about running this place is wrong."

Maya sat down across from him, her expression understanding but firm. "That's what happens when you ignore the numbers for so long. It doesn't mean you've failed, Alex. It just means we need to adjust."

"Adjust?" Alex muttered, rubbing his temples. "I feel like we need to tear the whole thing down and start over."

Maya leaned forward. "You don't have to start over. You just need to rebuild the foundation. We have the clients, we have the talent—we just don't have the systems to support the growth. That's what Jordan was getting at."

Alex nodded slowly, but the weight in his chest remained. Systems. Structure. They were the very things that had always made him feel trapped, like they were suffocating his creativity. But now, it was clear that not having them was the thing that had kept him in constant survival mode.

Maya stood and walked and grabbed the binder and flipped to the page where Jordan had written out the next steps: Gross Margin, COGS, Client Retention, Team Structure. "This is where we start?" she asked, pointing at the top of the list.

Alex responded "Yes. And it's not going to be easy, but if we fix these, the rest will follow."

Maya smiled faintly. "This is EXACTLY what we need. You started to build something special when you launched the agency. But the bigger it got, the more you needed this place to run on its own. You've been trying to do everything yourself for too long. We need to give the business the structure it needs to survive without you fixing every crisis."

Alex closed his eyes. This business had always been his— his vision, his creation. Letting go, even just a little, felt like losing a part of himself. But deep down, he knew Maya was right. He couldn't keep this up. The late nights, the constant stress—it was eating him alive.

Maya turned back to him. "You've got to trust us, Alex. Trust the team, trust the systems Jordan's helping us build. You're not letting go of control—you're gaining control in a different way."

Alex nodded. Trusting the systems. That was Jordan's mantra, and it was starting to sink in. But it didn't make it any easier. The thought of relying on numbers instead of gut instinct felt foreign, like stepping into a world he didn't understand.

"Alright," Alex said finally, his voice quiet but determined. "Let's start with the gross margin. What do we need to do?"

Maya smiled, her eyes lighting up with relief. "Good. I'll work on analyzing our pricing and COGS. We'll start with the services and the CLIENTS that are underpriced and adjust from there. We'll also look at where we're wasting money—inefficiencies, contractors we don't need, vendors who are overcharging."

Alex nodded again, though the tightness in his chest remained. Rebuilding the foundation—that's what Jordan had said. And it was clear that if he wanted this agency to survive, he had no choice but to tear it down to the studs and rebuild.

A couple more weeks passed and Alex found himself alone in the office, staring at the lists Jordan had left behind. He had barely slept, his mind racing with thoughts of numbers, systems, and the overwhelming realization that everything he had avoided was now the key to saving his business.

He glanced at the spreadsheet Jordan had sent over. Facts, not feelings. It was all there in black and white—the cold, hard truth he had spent years avoiding. His gross margin was a disaster, his cost of goods sold was bleeding the company dry, and the churn rate was worse than he had imagined. Every number was a glaring reminder of how far off track he had gone.

He opened his email and saw a message from Maya. She had already started working on the team analysis, breaking down who was performing and who wasn't. He skimmed through the list and felt his stomach drop. There were names on there—people he had worked with for years—who weren't pulling their weight. People he had avoided confronting because it was easier to pretend everything was fine.

He leaned back in his chair, running a hand through his hair. I can't keep avoiding this. It was a mantra he had been repeating to himself for weeks, but now it felt more real than ever. Jordan had made it clear—if he wanted to save this agency, he had to stop reacting and start leading.

Hey, how's it going?" she asked happily.

Alex hesitated before replying. He didn't want to worry her, but he couldn't keep pretending everything was fine.

"It's tough. Jordan's got me looking at all the numbers, and the team... it's not good. I'm feeling pretty bad about myself if I'm honest." Alex said embarrassed.

Her response was almost immediate.

"I'm proud of you for facing it. I know it's hard, but you're doing the right thing."

"I appreciate that. You sure you still want to be married to me even though I'm a loser?"

Alex replied with a joking tone, but you could tell that in the bottom of his heart he was feeling really low.

"Fuck no. I'm already signed up for Tinder." she replied. As absurdly as his question was, she knew that she had to reply with something even more ridiculous.

"Listen, I don't care if the business falls apart and we have to live in a cardboard box, I'm not leaving you. And you will...WE WILL get through this. It's going to be ok babe." and she replied with the firm and loving tone that she knew Alex needed to hear.

Alex smiled faintly, her words offering a small comfort in the midst of the storm. He wasn't sure he believed it yet, but knowing Emily was in his corner made it feel a little less daunting.

He glanced at the clock. Maya would be back in an hour to review the progress they had made.

"Thank you Em. Maya's going to be here in a little bit so I have to go. I'll see you at home. I love you."

"I love you, too. More and more every day." and she hung up.

Alex sat across from Maya in the quiet, almost surgical order of her office. The blinds were half-drawn, casting stripes of light across a pile of printouts and the open screen of her laptop, where the agency's metrics laid bare a story he'd avoided reading for too long.

"Let's start with time tracking," Maya said, her tone as steady as a heartbeat. She turned her laptop toward him, the screen filled with numbers that represented the hours his team had dedicated to each client. Each row seemed to shout a hard truth he'd been dodging: they were undercharging for almost everything.

"Look at these legacy accounts," she continued, her voice soft but unyielding. "We're bleeding time here. You're pouring resources into clients who haven't adjusted their expectations—or their budgets—in years."

Alex clenched his fists under the table. He knew these clients too well; they'd been with him since the agency was little more than a fledgling idea, back when every dollar felt like a victory. Raising prices on them felt personal, like a betrayal of that early trust. "These are people who were here before anyone else believed in us," he murmured, barely meeting her gaze.

Maya sighed, and the weight of her sympathy softened the edge of her next words. "I get it. I know they're important to you. But Alex, they're demanding more resources than they're paying for. At this rate, they'll drain us dry."

Her tone sharpened as she pulled up a few examples. "For new clients, we'll need to raise prices by 25% across the board. For the rest? A 10% increase will help offset costs, but these five accounts here..." She tapped the screen with her pen, each name like a stab to Alex's gut. "They're well below our current rates, and they're exceeding the scope of work. We're talking 35 to 50% hikes for them."

Alex's pulse raced as he took in the rows of numbers. Each dollar they'd undercharged over the years wasn't just lost revenue—it was lost time, resources, and morale. But it was also the fear: what would these clients think? How could he justify the increases?

"They'll leave," he said quietly, almost to himself.

Maya's gaze held his, her eyes resolute. "Maybe. But would you rather lose them now, or lose the entire business slowly? Your loyalty to them could cost you everything else you've built."

The truth of her words landed with the force of a hammer, but he could already see it—a new structure built on a foundation of stability, not sentimentality. He nodded, swallowing the thick knot of worry in his throat. "You're right. We'll do it."

Maya nodded approvingly and clicked over to another file. "Now, expenses. We've got some easy wins here." She ran through a list of software subscriptions, underused tools, and outdated contracts. "We can free up a few thousand a month with minimal impact. And as we go through, we'll look at the SOPs. There's a lot of overlap we can cut. Jordan was right—this is an issue of clarity as much as cost."

Alex's tension eased slightly, a glimmer of hope peeking through the fog. If they could cut down costs and raise revenue, they'd finally have a chance to catch their breath. "We'll start tomorrow," he said, feeling the stirrings of real determination.

Maya glanced at him, her face serious but tinged with

approval. "Good. But there's one more thing we need to address." She flipped to a new tab, a list of team hours and performance metrics filling the screen. She highlighted two names, and Alex's stomach dropped.

"These two aren't pulling their weight," she said, not unkindly but with an edge of finality. "They claim they're working too much, but the numbers tell a different story. These two have become... well, dead weight. Personality hires who bring charm but drain productivity."

Alex swallowed, the knot in his stomach tightening. He'd hired these people partly because they were likable, the kind of people who kept things light in meetings, but their contributions were minimal. One of them was notorious for talking circles around tasks without completing them, and the other... the other was the first to complain of burnout, despite logging the fewest hours.

"Do we let them go?" he asked, his voice barely a whisper. He'd avoided this conversation for so long, choosing to ignore the mounting frustration these two brought to the rest of the team.

Maya's gaze didn't waver. "Yes. Not because they're bad people, but because they're dragging everyone else down. We have people who are stepping up, covering for them, burning themselves out. We have to put the business—and the rest of the team—first."

The weight of her words settled on him, but instead of despair, he felt a new clarity taking root. For the first time, he could see a way forward—a way to strengthen the agency, not by holding on but by letting go.

He turned to Maya, nodding slowly. "Alright. We cut the dead weight. Raise the prices. Trim the expenses." He let out a breath he hadn't realized he'd been holding. "No more running on instinct."

Maya offered him a small, encouraging smile, one that carried the force of a thousand promises. "You're making the right call. I'll start drafting the client communication templates, and we'll plan meetings to make this as smooth as possible."

In that moment, Alex felt a new kind of weight—a gravity that felt right, purposeful. The choices ahead wouldn't be easy, but for the first time, he knew he could make them. And as Maya gathered her notes and left him to the silence of his thoughts, he realized that this was the beginning of something he'd longed for: a business that was built on purpose, not just passion.

The stakes had never been higher, but for the first time, he didn't feel afraid.

Chapter 3: The Launch
Level 1
Guided by Jordan: Weekly Breakthroughs

Alex stepped into the conference room, the low hum of fluorescent lights buzzing overhead. The air felt thick with tension—a tension he couldn't shake. Today wasn't just about numbers or data. Today was about confronting the parts of his business he'd been avoiding for years. The checklist lay open on the table in front of him, glaring at him with its stark, empty boxes.

Maya entered, carrying stacks of papers under her arm. She caught his eye and gave him a small, reassuring nod. "We'll tackle this together, Alex," she said, her voice steady, but even she couldn't mask the concern that flickered behind her eyes.

Jordan arrived moments later, exuding the same air of quiet authority that always made Alex feel both comforted and unnerved. Jordan's calm confidence only served to underscore how out of control Alex felt.

"Alright," Jordan began, his eyes scanning the checklist, "Let's go through it, step by step."

As they sat down, Jordan slid the binder toward Alex, flipping it open to the Level 1 checklist: "Launching." The name itself felt like a slap in the face. His agency was supposed to be well beyond "launching," but Jordan had insisted they start from scratch.

"I know it feels remedial," Jordan said, as if reading Alex's mind. "But if these foundational gaps are missing, it doesn't matter how much you've built on top of them. It's like building a skyscraper on sand."

Alex's eyes scanned the first set of items, his heart sinking deeper with each one. Core Values? Vague. Mission Statement? Never clearly defined. He had always relied on instinct, assuming that the agency's values were understood without needing to be explicitly laid out.

"These are the things that should guide every decision you make," Jordan said, tapping the page. "Without them, you'll always be reacting to the next crisis instead of steering the ship in the right direction."

Alex looked down, the words blurring slightly as his anxiety spiked. What had he been doing all these years? He prided himself on being a visionary, but here, staring at this checklist, it was clear that vision without structure had led him astray.

Jordan didn't waste time. "Let's talk about the org chart.

Who's doing what? And more importantly—are they doing what they should be?"

Alex swallowed hard. The truth was, his team had grown in a disjointed way. People had taken on roles they weren't suited for. Lines of responsibility were blurred. Maya, his Operations Manager, was doing more than half the work that belonged to others, while some employees were coasting, hidden behind the chaos. It had been easier to ignore it.

"Well," Alex started, his voice tight, "Maya handles most of the operations—honestly, she's taken on more than she should, but... I haven't had the bandwidth to fix it."

Maya looked at Alex but said nothing, her silence louder than any words could have been.

Jordan didn't blink. "You've got talented people here, Alex. But you've also got inefficiencies that are killing you. We need clarity—who's responsible for what, and who needs to step up or step out."

"Step out?" Alex echoed, his stomach twisting. He had never heard that term before.

"Yes," Jordan said bluntly. "If someone isn't contributing to the vision, they need to go. No exceptions. We're not running a charity here. Our company is like a sports team. If I hire you to score touchdowns and you can't do that, then you're not going to be on the team for very long."

The cold, hard truth of it hung in the air. Alex felt a cold sweat break out. His instinct had always been to protect his team, to avoid the difficult conversations. But now, he realized that avoidance had led him to this point—staring at a crumbling foundation.

"Well, the good news is that we have already identified two individuals that we want to let go" Alex replied.

"Fantastic. That is a great start and will also set the tone for the rest of the team members that our company culture is one of accountability." Jordan answered.

"And your numbers," Jordan continued, turning the page to reveal financial data. He flipped to the profit and loss statement, and Alex winced, feeling the weight of what was coming.

"We already know that your Gross Margin is way lower than it should be." Jordan said, matter-of-factly. "You should be at 70%. Did you figure out how to get those numbers moving in the right direction?"

"That's what these stacks of papers are for" Maya answered proudly. " I have found several expenses we can cut, clients we can raise prices on, and ways to make us more efficient. I think we can get to 55% within the next few weeks without blinking".

Jordan smiled. "This is exactly what we need Maya. Great job. Alex, you cool with all of this?"

Alex nodded, "Absolutely. This is the most excited I've ever felt about numbers and expenses in my entire life. Especially when I"m not the one having to do it."

They all had a quick laugh and then settled back into their difficult conversation.

"And let's talk about your churn rate," Jordan continued, flipping to another chart. "You're losing 10% of your clients every month. That's a recipe for disaster."

"Clients come and go," Alex muttered, his voice defensive. "It's part of the business."

"It's part of a business," Jordan corrected, "but not one that's run properly. That's 10% of your revenue walking out the door every month. You're stuck in a constant loop of replacing lost clients instead of growing and it's also a sign that your product or product experience is not living up to expectations."

Maya finally spoke up. "We've noticed it, Alex. Clients aren't getting the consistency they need from us. There's been too much inconsistency in the lead generation, and it's affecting the quality of work."

Alex had always believed in the strength of his relationships with clients, believing that personal rapport would outweigh any operational shortcomings. But the numbers told a different story.

"How do we fix it?" Alex asked, his voice barely above a whisper.

"We start with communication," Jordan said firmly. "Proactive check-ins, transparency, and setting expectations early. Clients need to know you're on top of things—not reacting to problems after they happen."

"And internally?" Maya asked, her voice calm but purposeful. "We've got people who don't know what's expected of them."

Jordan nodded. "That's why the org chart matters. Clear roles, clear expectations. No more overlapping responsibilities. We'll create systems that make it impossible for people to hide."

Alex then replied "I'm starting to see why this checklist is so important. Let's rebuild the foundation of this company."

Jordan next tapped on the checklist, now updated with several high-priority items after he checked off all the items Alex and the team had unknowingly created over the years.

"You've got a lot of these checklist items done through luck, but these next steps are still missing and these are non-negotiable," he said, meeting Alex's gaze with a firm resolve. "They'll help create that missing foundation your agency needs to operate effectively and consistently, and they're the steps that'll get you out of this cycle of chaos."

He flipped the page to the first item, **Create Version 1 Standard Operating Procedures (SOPs) for Key Fulfillment and Deliverables.**

Operations
- [ ] Create Version 1 Standard Operating Procedures (SOPs) for key fulfillment/deliverables.
    - [ ] Websites (2nd Priority)
    - [ ] FB Ads (4th priority)
    - [ ] Google PPC (5th priority)
    - [ ] SEO (1st priority)
    - [ ] Digital Lead Management/GHL (3rd priority)
    - [ ] Social Media (Weekly Posting, copywriting, video editing) (6th priority)
    - [ ] Print (7th priority)

"Your fulfillment team needs consistency. SOPs will ensure they approach every task the same way each time, which is the only way you'll get consistent results and control COGS. This will minimize guesswork, cut waste, and make your client experience dependable."

Alex nodded, understanding that the lack of uniformity was bleeding his agency dry. The team had talent, but talent without a system was costing him time, money, and clients.

Jordan's finger moved down the page to the next item: **Establish Top 3-5 Product Success Metrics and How to Measure Them, and Set Up Version 1 of the Customer Journey through the First Year of the Client Relationship.**

- [ ] Establish Top 3-5 Product Success Metrics and How to Measure Them
    - [ ] Websites
    - [ ] FB Ads
        - [ ] Number of Leads Generated
        - [ ] Cost Per Lead
        - [ ] % of Leads to Consults 30-40% or better
    - [ ] Google PPC
        - [ ] Number of Leads Generated
        - [ ] Cost per lead
        - [ ] % of leads to consults 30-40% or better
    - [ ] SEO
    - [ ] Digital Lead Management/GHL
    - [ ] Social Media (Weekly Posting, copywriting, video editing)
    - [ ] Print

"These metrics are essential. They'll give you concrete numbers to evaluate the success of your services. Plus, a defined customer journey with feedback points along the way will help you track your clients' satisfaction and progress, which is exactly what you need to keep churn under control."

"Having these metrics will give us proof of the value we're delivering," Maya added. "If we're getting clients measurable results, we're a lot less likely to lose them."

Define Job Descriptions and Roles for Every Position, Jordan continued.

Executive Assistant

I. GENERAL INFORMATION.

Position title	Executive Assistant
Work area	CEO
Reports to	CEO
Positions directly supervised	NA
Positions indirectly supervised	NA
Date of creation	March 2023
Date of update	TBD

II. POSITION'S GOAL.

The person in this role is responsible for overseeing the CEO's schedule and communication channels, effectively prioritizing and managing emails and phone calls, as well as arranging important meetings and business events. Their exceptional organizational skills and attention to detail enable them to keep the CEO informed, prepared, and on track, ensuring smooth operations and optimal productivity.

III. KEY FUNCTIONS AND RESPONSIBILITIES.

"You can't afford any more blurred lines on responsibilities. Clear roles will allow you to evaluate performance accurately, make sure you've got the right people for the job, and manage expenses by identifying unnecessary roles or overlapping duties. This clarity doesn't just keep costs down; it keeps your team members performing at their best."

Alex felt a wave of relief at the idea of finally cutting through the fog that had developed around roles and responsibilities.

Next was **Differentiate from Competitors with a Clear Unique Selling Proposition (USP).**

Marketing and Sales
- ☐ Lead Generation and Sales Quotas Created for Each month
- ☐ 1 Ideal Customer Profile
- ☐ Differentiate from Competitors with a Clear USP.
- ☐ Develop 1 Lead Generation Channel (cold email, warm call, podcast, etc)

"Your clients need to understand, quickly and clearly, what sets you apart and what value you bring. A solid USP will strengthen your marketing efforts and lead generation. Clients won't stay if they don't see your value clearly."

Jordan flipped to the last few items, placing his finger under Marketing and Sales: **Develop 3 Packages for Services and Integrate These into the Sales Deck.**

"This is where we stop the chaos," he said. "You need these packages to streamline delivery. Offering a standard set of packages means your team will know exactly what to deliver, and your clients will understand exactly what they're getting. We're getting rid of the randomness that has made scaling so difficult."

Maya glanced at Alex with an encouraging look. "This will make our lives easier too, Alex. If everyone's on the same packages, we can ensure quality without constantly pivoting."

Finally, Jordan highlighted Create Version 1 of Your Signature System. "This is the process that defines how you get results for clients. It's your secret sauce, and when your clients understand you have a proven process, they'll be more likely to stay. This Signature System will strengthen your brand and marketing appeal while keeping fulfillment straightforward."

Alex scanned the list, feeling the weight of the steps ahead. But for the first time, he also felt a tangible sense of relief. This wasn't just about "fixing" things; this was about building the structure that would finally give his agency the stability he'd been chasing.

Pushing Back on Structure: Resistance and Growth

Alex arrived at the office the next morning, the checklist from Jordan still on his mind, a reminder of all the steps he'd overlooked or dismissed. Standing at his desk, he picked up the thick folder, flipping it open to the first page. Each box felt like a reminder of everything he hadn't done—a silent but persistent nudge toward change. Instead of irritation, though, he felt a strange sense of dread mingling with curiosity.

Maybe this isn't about slowing down, he thought, but about clearing a path forward.

He poured himself a coffee, watching the steam rise, gathering his thoughts. In his early days, he'd relied on instinct, taking risks, and improvising through challenges. Now, he couldn't ignore the fact that his usual approach wasn't working anymore. His agency had grown beyond his ability to handle it by instinct alone. The chaotic mix of client work, team issues, and mounting pressure felt like it was spiraling out of control. He'd thought he could fix it all if he just kept pushing harder, but now he wondered if maybe Jordan's way—structured, intentional—wasn't the enemy of progress he'd assumed.

A light knock on his office door interrupted his thoughts. Maya walked in, holding her usual clipboard, but she was watching him closely. "Morning, Alex," she said. "Just wanted to check in. You seemed deep in thought."

Alex nodded, tapping the checklist lightly with his finger. "Just trying to wrap my head around this," he admitted. "Jordan wants us to slow down and set everything up properly, but part of me still feels like it's dragging us backwards."

Maya glanced at the open checklist and then met his eyes with a steady gaze. "I get it. But we've been going in circles for months. Maybe... it's time we try a different approach?"

Alex looked down, gathering his thoughts. "I just... I built this business by moving fast, trusting my gut. If I'd been following a rigid plan back then, I don't know if we would've gotten this far."

Maya listened, her expression compassionate. "But we're not where we were back then, Alex. Things have grown, and the agency's needs are different. You're running an entire team now—not just you and a few contractors anymore. This checklist isn't meant to slow you down; it's here to give you space to breathe and delegate."

He considered her words, eyes drifting back to the checklist. Despite his doubts, he couldn't deny the exhaustion he felt. It had been creeping up on him for months, and it was harder and harder to ignore. "I'll try," he said finally, the words a tentative commitment.

Maya nodded, her voice encouraging. "We're here to help. You don't have to do it all alone."

After she left, Alex's gaze returned to the checklist, skimming over the next action items: setting up standard operating procedures, establishing success metrics, and defining roles for everyone on the team. Each task seemed monumental, but for the first time, he saw how each box checked off could build a foundation that didn't depend on him carrying it all on his own.

As the day wore on, however, he couldn't shake the pull to handle things his way. A client project hit a roadblock, and his instincts kicked in—he stepped in immediately, attempting to solve the issue directly. But the chaos only deepened. By mid-afternoon, deadlines were slipping, and frustrations were rising among his team. It was clear that without the structure, things were unraveling.

Maya caught him in the hallway as he moved from one task to the next, her expression calm but direct. "Alex, the team's struggling," she said. "They're not clear on the priorities right now. We need to take a step back and look at that checklist Jordan gave us—it's supposed to prevent these kinds of issues. Right now, we're skipping steps."

Alex exhaled, feeling the truth of her words sink in. He'd avoided parts of the structure, hoping to make it work his way, but Maya was right. The team needed clarity, and so did he.

He gave her a resigned nod. "Alright," he said, with more conviction than he'd felt all day. "Let's regroup. No more skipping steps. We're doing this right."

The Power of Team: Stepping Up Together

It was a Friday morning, and Alex was not at his desk for once. He was still at home, lingering in the kitchen, staring absentmindedly at his coffee cup. Emily was rushing around getting ready for her day, but Alex seemed frozen in place. The weight of the previous conversations with Jordan was pressing down on him, a constant, uncomfortable reminder that he was still resisting the structured change his business desperately needed.

"You okay?" Emily asked as she zipped up her jacket.

Alex blinked, pulling himself back to the present. "Yeah,

just... thinking."

Emily paused, looking at him carefully. "Thinking or overthinking?"

"Maybe both," he muttered, rubbing his temples. "Jordan's pushing for all this structure, and I get it. But it's... hard. I built this place with my gut, you know? And now it feels like everything I did was wrong."

Emily stepped closer, placing a hand on his arm. "You didn't do it wrong. You were a different person back then and your business is in a different place now. You always said you wanted this business to grow, right? This is what that looks like. Growth is uncomfortable."

Alex sighed, knowing she was right. He had always imagined growth as something smooth and exciting, but this—this felt like wading through mud.

"Thanks," he said, offering a half-smile. "I just wish I didn't feel so... out of control."

"You're not out of control. You're just not in control of everything anymore, and that's different," she said, kissing him on the cheek before heading toward the door. "You'll figure it out."

Alex watched her leave and, with a deep breath, grabbed his jacket. He recognized how much he needed her. She was the one thing in his life that helped him maintain his equilibrium in life. For the first time, in a long time, he remembered that wealth is much more than just the numbers in his bank account. It was his relationship with his wife. And with that, it was time to head to the office.

By the time Alex arrived at the office, the hum of focused activity was already underway. He stopped at the doorway, watching his team for a moment. Maya was speaking quietly with Rebecca, her expression serious but confident. Ethan was on a call with a client, handling it smoothly. For a moment, Alex wondered if his constant presence had been more of a hindrance than a help.

Maya spotted him and waved, signaling for him to join her and Rebecca.

"Morning, Alex," Maya greeted him. Her tone was warm, but he sensed the undercurrent of focus and purpose. "We've started working on the items from Jordan's checklist. Rebecca's helping to set up Standard Operating Procedures for key deliverables, especially in fulfillment. We want everyone to follow the same approach, so there's consistency in what we deliver."

Rebecca gave a nod, her quiet confidence bolstering Maya's words. "We're focusing on creating basic SOPs that everyone can understand and use," she explained. "It's going to make things smoother for the whole team, and it'll save us a lot of time correcting mistakes. I was shocked at how differently the team was doing things, but now that won't be an issue anymore."

Alex listened, surprised by how organized they sounded. "That makes sense. SOPs... that's something we never got around to," he admitted, feeling a twinge of guilt at the oversight.

Maya continued, "We've also started looking into the top success metrics for our core services. Ethan is gathering data on client feedback to make sure we're tracking the right outcomes. Once we have a system for measuring these metrics, it'll help us know where we stand with every client."

Ethan, having wrapped up his call, walked over to join them. "We need to figure out exactly what metrics show client success and satisfaction. I've been noticing that we've lost some clients not because of our work, but because they didn't see how we were helping them reach their goals. It's something that can be fixed with the right metrics."

Alex nodded, recognizing how the lack of measurable results had led to client churn. "If we're getting them real, visible results, they're less likely to leave us. It sounds like this will help with churn, too."

"Exactly," Ethan replied. "And with those metrics in place,

we can make our results more tangible to clients."

They moved into the conference room, where David, the Finance lead, joined them. He opened his notebook, glancing at Maya before speaking. "We've also started defining roles and responsibilities more clearly. Right now, too many people are taking on overlapping tasks, and it's costing us time and money."

Rebecca nodded. "I'm handling HR tasks that should be with department heads. If we can clarify everyone's responsibilities, it'll free me up to focus on bigger priorities."

Maya chimed in, "The new org chart defines who's accountable for what and who's responsible for key outcomes. Alex, we know you've been the go-to problem solver, but it's time for everyone to own their roles."

Alex felt a wave of understanding wash over him. His role as the 'fixer' was holding everyone back, himself included. "Alright," he agreed, his voice a little tighter than usual. "Let's make it happen. I'll review the org chart later today and give you my input."

<hr />

A couple weeks later Alex joined the team in the conference room to review the checklist progress. They'd set up several key initiatives, and Alex could see the change happening. They were no longer waiting for him to approve every step; they were making decisions and owning their parts of the business. He felt a mix of pride and unease.

"David, where are we on defining our Unique Selling Proposition?" Alex asked, knowing that Jordan had stressed the importance of differentiating themselves from competitors.

David gave a slight nod. "We've started. Our goal is to craft a clear message that clients will understand instantly: what we do, why we're different, and how we deliver results. It's something we need to communicate across the board, especially in our sales materials."

Maya jumped in, "We're also working on standardizing our service offerings into three core packages. This way, there's consistency in delivery, and we can avoid the chaos of custom requests. Once we integrate this into our sales deck, clients will have a clearer idea of what they're getting, and the team will know exactly what to deliver."

Alex was getting excited now. With clear packages, they wouldn't be scrambling to deliver on every custom demand. "This should definitely help us focus and stay consistent," he agreed.

By the end of the day, after most of the team had left, Alex sat alone in his office, looking over the Job Descriptions, SOP drafts, Unique Selling Proposition, and the new Client Packages. They were simple, clear, and exactly what Jordan had been pushing him toward. It was unsettling how much they had achieved without him having to guide each step. In a strange way, he felt more like a spectator than the driving force.

A knock at his door broke through his thoughts. Maya stepped in, closing the door behind her. "Hey, just checking in before I head out," she said.

"Thanks, Maya. I appreciate everything you and the team are doing. I think... I needed this push more than I realized," Alex admitted, his voice softer than usual.

Maya gave him a warm smile. "I know it's hard to step back, Alex. But by letting us take on these roles, you're not losing control—you're building a stronger foundation for all of us. We're ready to help carry the weight."

Alex nodded, staring at the checklist once more. He was learning that leadership didn't mean controlling every aspect of the business but empowering his team to lead with him. It wasn't easy, but seeing the team step up made him believe it might be worth it.

Completing Level 1: Victory in Small Steps

It was early, and the sun was just beginning to rise, casting a warm glow through the kitchen window. Alex leaned against the counter, cradling a cup of coffee, his thoughts running as fast as the caffeine kicking into his system. The house was quiet—Emily had already left for a morning yoga class—and in the stillness, he found himself contemplating the culmination of weeks of effort on the Level 1 checklist. Today, they'd finally present everything to Jordan, and he hoped the progress they'd made would meet Jordan's standards.

The checklist sat open next to him on the kitchen island, each item carefully checked off. The team had addressed each one, and with every box crossed out, Alex felt the weight of his earlier doubts ease, replaced with a cautious optimism. Today, they'd show Jordan the finished work.

After scarfing down a quick breakfast, Alex took a walk to the office. The cool, crisp air was refreshing, cutting through the remnants of sleep and nerves. He pulled out his phone and sent a message to Maya: "Final team huddle at 9. Let's get ready to show Jordan we're done." They'd all worked hard, and it was time to pull it all together.

The Team's Final Push

As Alex entered the office, Maya met him at the door, clipboard in hand. "Morning. The team's ready for the 9 a.m. meeting. Everyone's wrapping up the last details on the checklist items, and I've pulled together a summary for Jordan's review."

"Perfect," Alex replied, glancing around the office. It felt different—more organized, more purposeful. Jordan's guidance and the new systems had brought a sense of calm focus to the chaos that once defined their workdays.

By 9 a.m., the team gathered in the conference room. The checklist was projected on the screen, every item noted and checked off.

Maya began the meeting. "Alright, everyone, we're here to review and confirm the final checklist items. After weeks of refining processes, implementing new systems, and adjusting to our roles, we're ready to present everything to Jordan."

Each team member outlined their progress, confirming they had addressed each item Jordan had pushed them to complete. Maya kicked off with the Standard Operating Procedures. "We've documented Version 1 of the SOPs for each of our core deliverables. These processes are now standardized across the team, ensuring every client receives consistent results."

Rebecca nodded as she added, "With these SOPs, we can now onboard new team members more efficiently. Everyone knows the steps, and we've reduced the errors that used to result from personal variations. This structure is going to save us so much time."

Next, Ethan moved to the success metrics and client journey. "We've finalized the top three metrics to define client success: Number of Leads Generated, Completion Times, and QA revisions. We're also rolling out a formal one-year client journey map, with set check-ins every week for the first 2 months and then once a month after that. This will allow us to capture feedback early and address any issues. This should directly address the churn we've been experiencing."

Alex chimed in, impressed. "Keeping clients engaged long-term should help reduce that monthly churn we were struggling with. I love this because we've never been this proactive before."

Ethan smiled, the confidence in his work showing through. "Exactly, Alex. Proactive communication is our best tool here. Essentially we are now preventing "messes" instead of cleaning them up after the fact."

David took the floor next, explaining the clarity achieved in the revised roles and responsibilities. "With our updated org chart, each role is tied to specific performance metrics, and everyone knows who's responsible for what. By clarifying these roles, we've eliminated overlapping tasks and are streamlining communication. Each team member now has a measurable impact on our results."

Rebecca added, "We're also updating everyone's job descriptions. It's clearer than ever what's expected in each role, which should improve accountability and reduce unnecessary back-and-forths."

The team then reviewed the next item: defining their Unique Selling Proposition (USP). David took the lead, summarizing their progress. "Our new USP highlights our unique approach to client engagement and results-driven services. It's front and center in all our new sales materials, emphasizing what differentiates us and reinforcing the value clients receive."

Sophia, who had helped develop the USP, added, "We've already integrated it into our pitch deck. It's making it much easier for prospective clients to understand why they should choose us over the competition."

Finally, Maya detailed the newly organized service packages. "We've created three core service packages, each one solving our client needs, just at different budgets and speeds. These packages bring clarity for both our clients and team, allowing us to streamline operations and avoid overextending ourselves on custom requests. The packages are now part of our sales deck, so prospects know exactly what they're getting—and so does our team."

Alex leaned back, feeling the significance of the progress they'd made. "We're ready," he said, looking at each of them. They'd tackled every item on Jordan's checklist and transformed the agency's foundation. Now, it was time to show Jordan.

Presenting to Jordan: A Proud Moment

Later that afternoon, they gathered in the conference room, ready for Jordan's arrival. Each team member had a summary of their updates, prepared to present their completed work. When Jordan entered, he looked around, noting the air of confidence in the room.

"Alright, show me what you've done," he said, his voice steady and encouraging.

Maya began the presentation, walking Jordan through each item. As they presented, Jordan's expression shifted from attentive to genuinely pleased. When Maya described the SOPs, he nodded approvingly. "This structure will save you so much time and hassle," he said.

Ethan outlined the client success metrics and the client journey, which earned a smile from Jordan. "Retention is all about making clients feel valued and consistently delivering results. You're on the right path here."

As they moved to roles and responsibilities, David shared the new org chart. "Everyone now knows their responsibilities and how their roles contribute to the agency's overall success."

Jordan leaned forward, visibly impressed. "This clarity in roles will make a huge difference. People need to know their purpose and accountability."

Sophia presented the USP and the service packages, detailing how they'd already been integrated into the sales deck. "These packages are designed to simplify both our sales process and our service delivery. Our team can focus on what they do best without reinventing the wheel for every client."

Finally, Alex took a deep breath and summarized the completed checklist, ticking off each item. "This team has stepped up in ways I hadn't anticipated. They didn't just get through the list—they owned it. I'm grateful for their dedication."

Jordan looked around the room, a rare smile breaking

his usual reserve. "I'm proud of all of you. You've put in the work, and it shows. Completing Level 1 isn't just about following instructions; it's about creating a strong foundation. You've built something that can grow."

The team exchanged glances, their pride evident. Alex felt a surge of relief and satisfaction. They'd done it. They were finally ready for the next stage.

A Moment of Reflection

That evening, Alex sat with Emily on their back porch, a glass of wine in hand. The stars dotted the sky, and a gentle breeze ruffled the leaves. For the first time in a long time, he felt at ease. They had completed Level 1 and even though Alex doubted the process at times, he now realized all the little details that were missing and ultimately prevented his agency from reaching its potential. For the first time in his professional career, Alex saw that he had a real foundation for the agency's future.

"We finished Level 1 today," he told her quietly.

Emily smiled, leaning her head on his shoulder. "I'm proud of you. I know it hasn't been easy."

Alex nodded, feeling the truth of her words. They had worked hard, but there was still so much to do. The sense of accomplishment was tempered by the understanding that this was just the beginning.

"We've got a long way to go," Alex said, his voice steady.

"But we're getting there."

Emily looked up at him, her eyes filled with love and encouragement. "One step at a time."

Alex took a deep breath, feeling the weight of the day lift. He wasn't running from the chaos anymore. He was facing it, step by step, with a strong team behind him.

One step at a time.

And What Might Your Marketing Agency Need for Growth?

If you're struggling to scale your agency and feel like you're running in circles, then you're not alone. Many agency owners face roadblocks when it comes to growth, often unsure where to start.

But what if you had a proven checklist that guides you through every critical step of scaling?
Inside this chapter, our character Alex talks about the very same Agency Growth Checklists that help him break free from the daily grind and build a business that practically runs itself.

Want to see the exact checklist that took him from chaos to control?
With actionable steps on everything from optimizing client fulfillment to refining team operations and boosting profit margins, this checklist can be yours.

Visit **www.factsnotfeelingsbook.com/checklists** and download all of our checklists for free and get the structure your agency needs to thrive.

Chapter 4: Validation Stage

Level 2

Progress Check: A Candid Review

Alex sat in his usual spot at the conference table, waiting as Jordan flipped through the completed checklist items from Level 1 one more time. The faint hum of the office seemed to amplify the tension in the room, though Alex tried not to let it show. He glanced down at his notebook, mentally preparing for whatever additional challenge or question Jordan would lay out next.

Jordan finally looked up, giving Alex an approving nod. "Six weeks, Alex. You and the team have done impressive work here. You've laid the groundwork, and that's no small feat."

A flicker of pride crossed Alex's face. "Thanks, Jordan. The team really pulled together—Maya, Ethan, everyone. I can feel the difference already."

Jordan leaned forward, his gaze sharp but encouraging. "Good. You should feel the difference. But remember, this is just the start. We've patched up a lot of the foundational gaps, but Level 2—the Validating phase—is where things get more serious. This is where we're going to see if our business can stand up to the pressure."

Alex nodded, swallowing hard. He knew better than to expect a smooth ride.

Jordan picked up the "Validating" checklist and skimmed through the items, flipping to the marked pages. "The Validating stage is where we are essentially proving to OURSELVES that our foundation is solid and we start to evolve our "game" so to speak. Again, you have a lot of this in place just by luck, but here's where you'll need to focus your energy next. We're going to address your operations, marketing, customer experience, and financial systems at a deeper level."

He paused, pointing to the first item. "First up: Develop and Document Scalable SOPs for Consistent Service Delivery. At this point, the agency needs SOPs for every single deliverable, end-to-end, in a way that runs without your direct involvement. The team can take over execution, but we need the systems that make consistency non-negotiable. This is going to be a push for you to step back even more."

Alex let out a breath. "Got it. SOPs are priority."

Jordan nodded. "Next, you'll need to Hire Project Managers. People who can own projects fully and are positioned for growth. They'll ease the operational strain on you and start taking on real accountability. Look for individuals who bring maturity and can handle autonomy."

Jordan paused, letting it sink in. Alex jotted down a few notes, nodding as he did.

"On the marketing side," Jordan continued, "you're refining your niche, packages, and product-market fit. The goal is to hit a 20% close rate minimum. You'll also need to implement a second lead generation channel— one that complements your primary lead source, which are referrals, but gives us another way to bring in new clients. This will take you beyond relying on just one marketing channel and improve your acquisition predictability."

Alex felt the pressure building. "Two lead channels— understood. I always hated relying on just one thing because if that thing doesn't work then the new customer flow dies."

"Then," Jordan continued, "we're going to enhance your Customer Experience. That means a more structured customer journey. Every client should have a consistent experience that leads to the results they're looking for. You'll aim for 3% churn, and that comes through clear, intentional communication."

Alex shifted, already envisioning the workload. "Refining the journey, making it seamless. Got it."

Jordan's expression softened slightly. "Also, upsells and

referrals. I know you've been delivering value, but now the team needs to engage clients with specific referral and upsell goals. It's about maximizing the lifetime value of every client relationship."

Alex exhaled slowly. "I can see the logic. It's a gap we haven't fully explored."

Finally, Jordan's tone became serious. "And, of course, financial tracking. Optimizing pricing to hit a 65-70% gross profit is essential. Any adjustments we can make to your pricing model will drive more profit, but you'll need to monitor financial KPIs every month. Revenue, profit, CAC—all of it needs to be consistently tracked."

Jordan paused, letting the weight of it all sink in. "This is about creating a business that's going to support you, not just demand from you. As you work through this checklist, you'll also be taking on CEO leadership training. You're aligning on one niche, validating results, and stepping back even further to let the team drive the process."

Alex met Jordan's gaze, feeling the scope of the work ahead but also the stability that came with a clearer path. "Alright. Six weeks for Level 1... I think we can knock this out in two to three months."

Jordan nodded, a slight smile breaking his serious expression. "Time for you to get to work."

Handing Over Control: The Shift Begins

The soft hum of the coffee shop buzzed around Alex as he stared out the window, watching cars drift by. It was an unusual spot for a Monday morning meeting, but Maya had suggested a change of scenery—a way to clear their heads and step out of the grind. The smell of roasted beans and the clinking of cups filled the air, grounding him in a way the office hadn't been able to lately.

He checked his phone again, glancing at the time. The team was on their way, and today was the first day they'd be diving into the Level 2 checklist—Validating. After his meeting with Jordan, Alex was excited for the next step and he knew this was going to be a new level of challenge. This checklist required systems that could support real growth, and he knew the team would need to be on board and fully engaged. For once, he wasn't leading the charge on all the items; his role was to support them as they took ownership.

The door swung open, and Maya walked in, followed by Ethan, Rebecca, and David. Each carried a laptop or binder, their faces serious but determined. The sight of them all together, as a cohesive unit, brought an odd sense of relief and pride that Alex hadn't expected.

Maya sat down first, offering Alex a nod. "Ready to knock this out?"

He nodded, even though his mind was racing. "Let's do it."

As they settled in, Maya pulled up the Level 2 checklist on her tablet, projecting it onto the small screen behind them. The words "Validating Systems" glowed softly, and Alex felt his gut twist slightly. This wasn't just another list of tasks; this was about creating structure that could function without his constant oversight.

"Alright, team," Maya began, her voice calm with the quiet authority she had taken on recently. "We've got some clear targets now. Jordan outlined the next items we need to tackle, and each one will require us to be fully committed. This checklist is about validating everything we've built so far and making it scalable. No shortcuts this time." She glanced at Alex, her expression supportive yet firm.

Alex looked down, half-embarrassed, half-grateful. She wasn't wrong. He had been tempted, more than once, to dive headfirst into the next big idea without fully finishing what they had started. But that wasn't an option anymore.

"Ethan," Maya continued, "you're going to take the lead on refining the customer journey and managing churn. We've had clients leaving because they didn't feel the connection after we onboarded them. We need proactive touchpoints that make clients feel valued and engaged—before issues arise."

Ethan leaned back, a small smile on his face. "I've already got some ideas. We need to establish quarterly check-ins and feedback sessions for all clients, not just the high-value ones. This will keep our finger on the pulse before things go south."

Maya nodded, pleased. She turned to the team, her focus shifting toward fulfillment. "Alright, one of our big checklist items is to develop and document scalable SOPs for consistent service delivery. These need to be rock-solid, set up to run without Alex's involvement. We're looking for clarity, efficiency, and consistency across the board, no exceptions."

Maya paused, glancing at her tablet, before addressing the room. "I'll take this on and work with each department to standardize our processes. We need a clean, repeatable approach for every service we offer, and I want these SOPs finalized by the end of the month."

Rebecca, who had been listening intently, spoke up, her usual quiet demeanor edged with resolve. "I'll support you, Maya, by making sure we have updated job descriptions that align with these new processes. We need everyone in roles that match their strengths, and each person should be crystal clear on responsibilities and expectations."

Maya nodded in agreement, appreciating Rebecca's commitment to refining the team structure alongside the SOPs. "Perfect, Rebecca. We'll work together to ensure these roles are defined and aligned with our objectives. Our goal is to build a framework that runs smoothly without constant oversight from Alex."

Alex watched them work through the plan, a sense of relief washing over him. They were all stepping up, each in their area of expertise, moving the agency forward in ways he'd been struggling to accomplish alone.

Maya then turned to David, the finance lead. "David, we need to tighten up on financial tracking and pricing. Jordan made it clear we're aiming for 65-70% gross profit. Can you handle the monthly tracking and identify where we need to optimize?"

David, always the numbers guy, nodded, already jotting down notes. "I'll create a monthly dashboard for our key metrics. That'll give us a real-time view on revenue, profit, CAC, and churn. And I'll work with you, Alex, on refining our pricing model so we hit those targets."

Alex felt the familiar pull of impatience stirring inside him again, the urge to jump in, fix things, solve problems before they even became problems. But he pushed it down. They had this. He had to believe that they had this.

"Finally," Sophia, the head of Marketing and Sales, began, "we're going to implement a second lead generation channel to diversify our client sources. Right now, we're too reliant on referrals alone. Alex, your insights on our ideal client profile will be crucial, but I'll lead the charge on setting up the campaign, refining our messaging, and targeting."

Alex nodded, feeling reassured by the team's clear alignment on roles. This checklist felt different—more complex, yes, but also more organized. They were focused, and as he looked at each of them, he felt a new sense of confidence. Jordan's words echoed in his mind: "You can't grow this thing on your own. You need them to lead just as much as you do."

Still, the tension lingered. Alex's mind flashed back to earlier that morning, standing at the kitchen counter, coffee in hand, staring at the unmade bed. What if this doesn't work? The thought had come, uninvited, creeping in the way it always did. Doubt. Fear. The feeling that if he wasn't in control, the entire agency would unravel.

Emily had walked in, seeing the look on his face. "You're worried again," she had said, her voice soft, but knowing. "Let them do their job, Alex. You don't have to carry it all."

She had kissed him on the cheek, and the kids had come barreling into the kitchen, breaking the moment. But her words had stuck with him, lingering like a challenge. Let them do their job.

Now, sitting here in this café, watching his team take ownership of the checklist, Alex finally understood what she had meant. They didn't need him to fix everything. They needed him to lead.

And leadership wasn't about doing all the work—it was about making sure the work got done, by the right people, in the right way.

Maya glanced over at him, catching his eye. "We've got this, Alex," she said, her voice steady. "Trust us."

Alex took a breath and nodded. "I do."

Trust and Empowerment: The Team Rises

The early morning sun filtered through the blinds in Alex's office, casting long shadows across the desk. It was quiet in the house, a rare stillness before the day fully woke up. Alex, sipping his coffee, stared at the client reports on his laptop, tapping his foot anxiously. Everything in him was pulling to jump in, to take charge, to fix the client retention problem that had surfaced last night. But Jordan's words echoed in his mind, calm yet firm: "Trust the process, trust your team."

He had been through this before—this endless cycle of wanting to dive in, wanting to solve the problem himself, yet knowing deep down that wasn't sustainable. His head told him to delegate, to let Maya and the others handle it. But his gut, the thing that had driven his business for years, was screaming at him to act. Could he really just sit back?

Alex closed the laptop, leaning back in his chair as his phone buzzed. A message from Maya popped up on the screen.

Maya: "Morning! Ethan and I are handling the client issue. We've got the new retention strategies in place, and I'll update you later. Don't worry—we're on it."

He exhaled sharply, tension creeping up his spine. She had it under control. Logically, he knew that. Maya had proven time and time again that she was more than capable.

But there it was again—that pull to get involved, to take charge.

Letting go felt unnatural.

He grabbed his phone, hesitating for a moment before typing out a reply.

"Got it. Keep me posted."

As soon as he hit send, the unease set in. His thumb hovered over the phone, almost reflexively opening the email app. But then he forced himself to stop. No. He had to let them handle this. This was the whole point, wasn't it? Trust the systems, trust the people. He had to step back.

At the office, Maya moved with purpose. She and Ethan had set up a meeting with the client first thing in the morning. The client had expressed concerns about a lack of communication, but Maya wasn't worried. They'd been building better client feedback loops for weeks, and now it was time to show the results.

Ethan, sitting across from her, glanced up from his notes. "Think Alex is going to check in mid-call?" he asked, half-joking, but Maya could see the real question behind it. Alex had a habit of swooping in at the last minute, tweaking their plans, leaving them scrambling to adjust.

"He won't," Maya said, with more confidence than she felt. "We've got this."

Ethan nodded, but there was a flicker of doubt in his eyes. Alex had always been the glue holding everything together, but they needed to prove that the agency could thrive without him micromanaging. That's what Jordan had emphasized in their last team meeting—trust the system, trust the team. Maya repeated that mantra in her mind as the Zoom call connected.

Back at his desk, Alex felt restless. He opened and closed his email several times, forcing himself to stay out of the client issue. He glanced at the checklist Jordan had given him, still half-complete. He had other tasks to focus on, but his mind kept drifting back to the meeting Maya and Ethan were having. Could they really pull it off without him?

His phone buzzed again. It was Jordan.

Jordan: "Check-in later today? Let's go over some of your numbers."

Alex stared at the message. Numbers. Facts. It was supposed to be straightforward, but why did it always feel so damn complicated? He was supposed to lead based on the data, not his gut. But wasn't his gut what had gotten him this far?

He shot back a quick reply, agreeing to the check-in. His anxiety was a low hum beneath the surface now, nagging at him. What if Maya needed him? What if the client wasn't happy?

But then a thought struck him. Maybe that was the problem. He always assumed the worst, that the team couldn't handle it without him. Maybe his need to fix things wasn't about protecting the agency—it was about his own fear of letting go.

Meanwhile, Maya was wrapping up the client call. It hadn't been easy—there were tough questions about past missteps, but Maya remained composed. She walked the client through the new communication process, explaining how they'd proactively provide updates moving forward. Ethan chimed in with data, showing how their lead generation strategy had already increased client engagement by 15% over the last quarter.

The client's tone shifted. "This is what I've been wanting to see," he said, a hint of satisfaction in his voice. "It feels like we're moving in the right direction."

Maya smiled, feeling the weight lift off her shoulders. They had done it. Without Alex stepping in. She typed a quick message to him once the call ended.

Maya: "Client's happy. We're good to go."

When Alex saw Maya's message, a mix of relief and frustration surged through him. They had done it—

without him. The realization hit harder than he expected. This was what he'd wanted, right? To trust the team, to step back. But why did it feel like he was being sidelined?

He didn't have long to dwell on it before his phone buzzed again. Jordan's face filled the screen as their Zoom call connected.

"Alex," Jordan greeted, his usual calm, direct tone. "How's everything going?"

"Good," Alex replied, though his voice was a little tighter than usual. "The team handled the client retention issue. Everything's fine."

Jordan nodded. "And how did that feel? Letting them handle it?"

Alex hesitated. "It felt... strange. I mean, it's great that they did it, but I feel like I should've been more involved."

Jordan leaned back, his face thoughtful. "That's normal. You've been the center of everything for so long, but that's not scalable. The more you step back, the more the agency can grow without relying on you for every decision."

"I know that," Alex said, rubbing his temples. "It's just... hard to let go."

Jordan's expression softened, but his words remained firm. "You have to trust them. This isn't about you doing everything. It's about building something that can run without you. That's what you're working towards, right?"

Alex nodded, the weight of Jordan's words settling over him. This was the whole point. He wasn't building a job for himself—he was building a business that could thrive without him.

Jordan shifted the conversation to the numbers, pulling up a dashboard with the client retention metrics. "Let's look at the data," he said. "You've improved your churn rate by an additional 3% in the last month. It's already down to 4% which is fantastic. Your lead generation has increased by 15%, and operational efficiency is up by 10%. That's not just a feeling, Alex. Those are the facts."

Alex stared at the screen. The numbers didn't lie. His team was performing. The systems were working. But it wasn't enough to just see the data—he had to believe in it.

"Trust the data," Jordan said, his voice cutting through the silence. "Trust your team."

Later that evening, Alex sat at his kitchen table, a glass of wine in hand, staring out the window as the last traces of daylight faded. Emily joined him, sitting across from him with a soft smile. She could sense something was on his mind.

"How did it go today?" she asked, her voice gentle.

"They handled it," Alex replied, his voice distant. "Maya and the team—they handled the client issue perfectly. I didn't need to step in."

Emily raised an eyebrow. "That's a good thing, right?"

Alex nodded, but there was hesitation in his eyes. "Yeah, it's what I've been working towards. It just... feels weird not being the one to fix things."

Emily reached across the table, placing her hand on his. "You've been the fixer for so long. But now you're leading. There's a difference."

Her words hit him in a way Jordan's hadn't. Alex squeezed her hand, the weight of the day finally settling in. He was shifting, growing. But growth wasn't always comfortable.

"You're right," Alex said, his voice steadier now. "I'm not the fixer anymore. I'm the leader."

And for the first time, he started to believe it.

The Second Milestone: Completing Level 2

The conference room buzzed with a quiet, focused energy, the whiteboard at the front covered with organized tasks— each item marking a final hurdle on their Level 2 checklist. The team sat around the table—Maya at the head, Rebecca flipping through HR files, David's laptop displaying a patchwork of financial projections, Ethan with a tablet, scrolling through client data, and Sophia, reviewing lead metrics.

Alex watched, absorbing the scene. For once, he wasn't leading the meeting. Maya had taken on that role, her calm, methodical voice guiding the discussion forward.

"Alright," Maya began, pointing to a note on the board. "Let's start with client journey refinement. Ethan, where are we with that?"

Ethan looked up, his expression focused. "We've implemented the monthly check-ins across all clients, and the new feedback forms are in place. We're already seeing an additional 4% improvement in client churn. Churn is already down to 4%. A few clients were hesitant at first to the monthly calls, but our team handled those individually, and now the feedback loop is solid."

Alex felt a flicker of pride. Just months ago, the idea of a formalized client journey had seemed unnecessary, a layer of detail he'd never believed they needed. Now, it was becoming a cornerstone of their fulfillment and customer experience.

"And on HR?" Maya prompted, switching gears effortlessly.

Rebecca leaned forward, her confidence growing. "We've finalized the evaluation process. Quarterly reviews now tie directly into our bonus structure and promotions, so accountability is crystal clear. Job roles are mapped out in detail—everyone understands what's expected."

"Good," Maya nodded. "This'll help cut down on the turnover that's been slowing us down."

Alex looked around the table, sensing the unity in each team member. They weren't just following orders; they were actively steering the agency forward.

Maya moved to the next item on the list. "Sophia, how's it going with finalizing our niche and implementing a second lead generation channel?"

Sophia adjusted in her seat, bringing up a report on her tablet. "We've refined our client packages to make our niche clearer, and that's started lifting our close rate. As for the new lead channel, I've set up a pilot for targeted ads to diversify our sources beyond referrals, and we're seeing promising engagement."

Alex blinked, impressed. This kind of multi-channel lead generation was something he'd always talked about but hadn't fully committed to. Sophia was making it happen.

"Anything left to refine?" Maya asked, though a glimmer of excitement showed in her eyes. The team was hitting their stride, and they all felt it.

Sophia replied, "We'll need a few more weeks of data to optimize targeting and messaging. Once we see conversion results, I'll adjust the strategy for our next campaign."

Maya nodded. "Perfect. Let's keep tracking and reviewing that weekly."

David, who had been listening intently, finally spoke up. "We've hit a snag with our contractor expenses."

Alex's pulse quickened as he leaned forward, feeling the old urge to step in. "What kind of snag?" he asked, his voice tighter than he intended.

David's fingers moved over the keyboard, pulling up a spreadsheet. "We're overspending on contractor costs. Several major accounts are costing more to manage than expected. If we can't renegotiate or cut costs, we're looking at a 5% margin drop this quarter."

The room went tense. Everyone knew the agency's financial health was still vulnerable after the heavy investments they'd made. A 5% margin drop could compromise the stability they'd worked hard to achieve.

Alex instinctively wanted to grab the phone and dive into negotiations himself. But when he met Maya's eyes, she held his gaze, her look saying: Let us handle it.

He took a breath, leaning back. "Alright. What's the plan?"

David didn't miss a beat. "I've identified three contracts where we're paying above market rate. I'll handle the renegotiations. If they don't adjust, I've lined up backup vendors we can transition to."

Alex hesitated but ultimately nodded. He'd hired David for his expertise. It was time to let him do his job.

"Good work," Alex said, feeling the tightness in his chest ease. "Keep me posted."

As the meeting wound down, each item on the checklist was reviewed, checked off, and celebrated. Maya presented the fully documented SOPs for service delivery, ensuring every step was structured to work without Alex's direct involvement. Ethan finalized the Client Journey, making upsell opportunities part of the process, and Rebecca confirmed that the refined HR processes were set and ready for the next hiring phase. Sophia reported that the second lead generation channel was now operational, and David's renegotiation plans were already in motion to protect their margins.

"That's it," Maya said, turning to Alex with a rare, satisfied smile. "The Level 2 checklist is complete."

A quiet moment settled over the room, the weight of their progress sinking in. For months, they had been meticulously rebuilding the agency from the ground up, tackling one foundational problem after another. And now, the work they'd put in was complete.

Alex looked around at his team—his team—and felt something shift. This wasn't just about him anymore. The team had made the agency what it was—not his gut instincts, but their dedication, discipline, and trust in one another.

He swallowed back a hint of emotion and addressed the team. "You all crushed it," he said, his voice thick with pride. "This agency wouldn't be where it is without each of you."

Maya gave him a small nod, a look of pride and mutual respect passing between them. They all knew how far they'd come.

Later that night, Alex sat in his living room, reflecting on the day. The Level 2 checklist was finally behind them and all that was left was to review everything with Jordan. Over the course of 2 months his team had stepped up in ways he hadn't imagined, proving that the agency could thrive without his constant input. The realization settled over him—he didn't need to be the center of everything.

Emily walked in, a glass of wine in hand, and took a seat next to him. Her eyes held that familiar, knowing look.

"You look different," she said, studying his face. "Lighter."

Alex chuckled softly. "Yeah, I guess I am."

"You finished the checklist?"

He nodded, the magnitude of the accomplishment weighing on him in a good way. "We did."

She smiled, leaning into him. "I told you—you're building something bigger than yourself."

Alex wrapped his arm around her, grateful for her steady presence. She was right as usual. For the first time, he felt like he was truly building a real business that didn't depend on him to survive but one that would thrive because of the systems and the people he had trusted.

Reflect and Refocus: A Fresh Perspective

This time Alex and Jordan decided to meet for dinner at one of the swankier restaurants in town. Alex always avoided places like this because deep down he never felt good enough to fit in with the crowd that frequented these establishments. And then on top of that he couldn't easily justify dropping $1000 on a meal for two people. However, Jordan insisted that they splurge a bit and conduct their next meeting in a nicer location.

The Level 2 checklist lay on the table in front of him, every box meticulously checked off. He stared at it for a moment, the weight of its completion hitting him harder than he'd expected.

For Alex, this wasn't just about finishing tasks. It was about stepping into a new version of himself, one that wasn't built on gut decisions and frantic action, but on something far more stable: trust in his team and in the systems they had built together. He picked up his pen, tracing the lines of the last item they'd completed—client feedback loops. It had been one of the toughest changes for him to accept, but now it was one of the most significant improvements they had made.

He could see it now. Clearer processes, fewer fires to put out, clients staying longer. A smooth machine, humming to the rhythm of data-driven systems, not emotional knee-jerk reactions. And it wasn't just the business that had changed. It was him. The relentless drive to control everything had loosened its grip. He had stepped back, and the team had stepped up.

Maya had been instrumental. Her leadership, calm but firm, had given the operations a new rhythm. Rebecca had formalized the HR systems, implementing reviews and training that made the team feel more supported and aligned. Ethan's communication with clients had brought retention rates up, and David's financial projections now provided them with a path to sustainable growth. They had built this together—each of them filling a critical role.

As Jordan reviewed everything that Alex and his team had put together, Alex took a deep breath, feeling a strange mix of pride and anxiety. They had completed Level 2, but was there something maybe that Jordan would see and make them start over?

Jordan leaned back from the table, his eyes narrowing slightly as if he could read Alex's thoughts through the screen. "So, how does it feel?" he asked, his voice steady but probing.

Alex paused. How did it feel? A lot of things, really—relief, pride, fear. "It feels... like we're finally getting somewhere," Alex admitted, his voice quieter than he'd expected. "The team's in sync, the systems are running. We've made real progress."

Jordan nodded slowly, but there was a flicker of something in his expression, something that told Alex this conversation wasn't going to be all congratulations. "That's good," Jordan said, his tone neutral. "But we're not done. Tell me, what's the one thing you're still afraid of?"

Alex blinked, thrown off by the question. Fear? He hadn't thought about it like that. But now, with the checklist complete, there was an uneasy silence underneath the satisfaction. "I guess... I'm afraid I'll slip back into old habits. That I'll get too comfortable and start ignoring the systems again, thinking I can do it all on instinct."

Jordan's eyes softened, but his voice stayed firm. "That's exactly what I was hoping you'd say. Because you will be tempted to fall back, especially when things get harder in the next phase. Scaling is going to challenge you in ways that this checklist never did."

Alex felt his stomach tighten. He had known this, somewhere deep down. Levels 1 and 2 were about plugging the leaks in a sinking ship. But Level 3? That was about sailing into uncharted waters, about growth and expansion. And it terrified him. Could he really let go enough to lead without micromanaging? Could he trust the systems and the team to keep the ship steady?

"We've got the data now," Alex said, trying to sound more confident than he felt. "Our retention rates are up, leads are more consistent, and our profit margins are finally stabilizing. It's all there in the numbers."

Jordan gave a small, approving nod. "Good. But numbers are only half the story. Don't forget that. The systems will work, yes, but only if you keep leading from the front. And leading doesn't mean controlling everything—it means having the vision to see the bigger picture."

Alex let those words sink in. Vision. That had always been his strength, hadn't it? Seeing possibilities others didn't, pushing the boundaries of what his agency could become. But Jordan was right—vision without structure had been his downfall. Now, he had both.

"We're moving into the Professionalizing phase next," Jordan continued, his voice steady but urgent. "And this is where you'll be tested the most. It's not about solving day-to-day problems anymore. It's about building something that can scale beyond you."

The gravity of Jordan's words weighed on Alex. They had caught up, they had fixed the leaks, but now they were about to face bigger challenges. Would the systems hold? Would the team thrive without his constant oversight?

Jordan must have sensed Alex's hesitation. "You've done the hard work, Alex. You've built the foundation. But now it's about thinking long-term. Are you ready to build something that can outlast you?"

Alex stared at his glass of Merlot, feeling the enormity of the question. Outlast him? He had always seen the agency as an extension of himself, but now... now it was becoming something bigger. Something that could operate without him someday.

"I am," Alex said, the words coming out stronger than he expected. "I'm ready."

Jordan leaned forward slightly, a small smile tugging at the corner of his mouth. "Good. Because this is just the beginning. You and your team did it. Great job. Let's get ready for Level 3."

And from that, the two finished the rest of their meal and reminisced as friends. Alex was able to relax and be himself knowing that he didn't need to feel inferior anymore to his friend and mentor.

Later that night, Alex sat on the back porch of his house, a little buzzed from the dinner with Jordan, while a cold breeze rustled through the trees. The stars were bright, scattered across the sky like tiny sparks of light. Emily was beside him, her hand resting gently on his. They hadn't spoken much since he got home, but the silence was comforting, peaceful.

"It's official. We're done with Level 2," Alex said quietly, staring out at the darkened yard. "Jordan gave us the thumbs up that we finished the checklist."

Emily smiled, squeezing his hand. "I knew you would."

Alex turned to look at her, taking in the soft curve of her smile, the way her eyes reflected the starlight. "It's more than just finishing the checklist, though. I feel like... I've changed. I'm not running everything by gut anymore. I'm trusting the team, the systems. And... I'm starting to trust myself."

Emily nodded, her gaze steady. "That's the real progress, Alex. The business is stronger, but so are you."

Alex leaned back, closing his eyes for a moment, letting her words sink in. Stronger. Maybe he was. Maybe this wasn't just about the agency—it was about him finally letting go of the need to control everything, to prove himself constantly.

He opened his eyes, staring up at the sky. The next phase was coming. Bigger challenges, bigger risks. But for the first time, he felt ready.

"You know," he said, glancing over at Emily, "Jordan asked me if I was ready to build something that could outlast me."

Emily raised an eyebrow, intrigued. "And what did you say?"

Alex smiled, a quiet confidence settling over him. "I said I was ready."

Emily grinned, leaning her head on his shoulder. "Good. Because I believe you."

And What Might Your Marketing Agency Need for Growth?

If you're struggling to scale your agency and feel like you're running in circles, then you're not alone. Many agency owners face roadblocks when it comes to growth, often unsure where to start.

But what if you had a proven checklist that guides you through every critical step of scaling?
Inside this chapter, our character Alex talks about the very same Agency Growth Checklists that help him break free from the daily grind and build a business that practically runs itself.

Want to see the exact checklist that took him from chaos to control?
With actionable steps on everything from optimizing client fulfillment to refining team operations and boosting profit margins, this checklist can be yours.

Visit **www.factsnotfeelingsbook.com/checklists** and download all of our checklists for free and get the structure your agency needs to thrive.

Chapter 5: Growing Pains
Level 3
Under Pressure: The Weight of Scaling

The afternoon light filtered into Alex's office, casting long, shifting shadows across his desk. He stared at the freshly printed Level 3 checklist, each item a stark reminder of the work ahead. The team had accomplished so much, yet this checklist seemed more daunting than anything they'd tackled so far. As he glanced down at the ordered timeline and ambitious deadlines Jordan had set, the words seemed to blur together. Outside, the hum of the city seemed louder—constant, insistent, like the agency's relentless demands.

His phone buzzed again, an alert from Ethan about a client issue. A fire to put out. The urge to dive in, to fix it, pulled at him. But he forced himself to resist, closing his eyes and taking a slow breath. Just as he reached for his laptop to review emails, the door to his office opened quietly. Maya walked in, tablet in hand, her movements precise, efficient, and deliberate.

"Alex," she said, setting her tablet on the table. "We need to talk about the operations review. There are still gaps we need to close before we can move forward."

He sighed, closing his laptop with a click. "I know, but every time we plug one hole, another springs open. We're growing, but I feel like I'm just barely keeping things together." He gestured at the Level 3 checklist, a slight frustration creeping into his voice. "And this? It feels like trying to keep a hundred balls in the air."

Maya leaned forward, her eyes steady. "That's exactly why Jordan set this checklist, Alex. He knew this phase would stretch us all. It's not about running a startup anymore; we're transitioning to a fully professionalized agency. That means trusting the systems even more, letting the team fill in the gaps, and taking the agency from sustainable to scalable."

Alex scanned the checklist. Each task had a deadline, each objective layered with new challenges. Jordan had laid out a sequence, urging them to complete these items in the next nine to eighteen months, each with a specific focus to ensure nothing slipped through the cracks.

Operations: First up was reviewing and expanding fulfillment capacity and developing scalable SOPs. Within three months, they needed to hit 70-85% team utilization rates and bring in middle management to ease the daily pressures. They'd need to implement a formal QA process and make sure all project timelines met the agency's new, more demanding standards.

Marketing & Sales: They needed to get a second lead generation channel running smoothly within two months to balance their over-reliance on referrals. By three months, Sophia would be responsible for developing a streamlined sales pipeline, all while pushing to improve their close rate to 20%. And with every sales interaction, they'd focus on refining the agency's unique selling proposition to help them break into new markets.

Customer Experience: Ethan would lead the charge on reducing churn by improving the client journey. He had one month to finalize this, with the goal of bringing churn down to 3%. Additional retention strategies, including loyalty programs and onboarding enhancements, would need to be solidified by the end of month two.

Finance & HR: David was tasked with refining financial tracking, setting up monthly reports on key metrics, and revising the pricing model to ensure a 65-70% gross profit margin. Additionally, they'd need two new leadership hires to handle scaling demands, with a goal to complete recruitment by month three.

CEO/Founder Development: Jordan had saved Alex's tasks for last. Over the next few months, Alex would need to shift even further away from daily operations, empowering the team to handle decision-making. He'd have to let go, aligning more with the agency's long-term vision and managing strategic risks to prepare for growth.

Maya's calm but steady voice broke through his thoughts. "I know this looks like a mountain, but we've done this before. Each piece is achievable—if we tackle it in order. Operations first, then Marketing and Sales, followed by Customer Experience, Finance, and HR, with you taking on leadership training alongside it all."

He looked at her, nodding slowly. She was right. They'd done it before, but he could feel the difference this time. This checklist wasn't just about plugging gaps; it was about building an agency that would outlast him. But with each task, the sense of chaos and urgency only seemed to grow.

"I hear you, Maya, and I know we need to follow the plan. It's just..." He gestured at the checklist. "It's overwhelming."

Maya nodded, giving him that look—one of gentle patience mixed with challenge. "That's the point, Alex. You're not supposed to do this alone. We're here to handle it with you."

She sat across from him, folding her hands over her tablet. "Let's start with fulfillment tracking and team utilization," she suggested. "It's the first deadline on Jordan's list. I'll get a report on where we are, and we can identify where to increase efficiency."

He glanced down at his checklist, finally picking up a pen and jotting down notes. "Alright. We'll start with fulfillment tracking," he agreed, his voice steadier.

They went over the rest of the list together, each item feeling slightly more manageable with her help. With each word, the tension in his shoulders eased a bit more, replaced by a growing resolve. Jordan's plan was ambitious, but it was the guide they needed.

For the first time that day, Alex felt a small spark of hope. They had a roadmap, a sequence, and a timeline. And he wasn't alone in following it.

Emily's New Perspective: A Conversation of Mutual Growth

Later that evening, after Alex's rough day at the office, he found himself sitting on the living room couch, lost in thought. The house was quiet, save for the distant hum of the dishwasher. Emily, ever perceptive, walked in, a cup of tea in her hands, and sat down beside him.

"You've got that look again," she said, nudging him with her elbow.

Alex offered a half-smile but didn't say much. He stared ahead, trying to make sense of the chaotic mess that his thoughts had become. Scaling the business felt like standing in the middle of a storm, and for the first time in a while, he wasn't sure he could weather it.

Emily studied him for a moment, setting her tea aside. "You're still doubting yourself, aren't you?"

He sighed. "Yeah. This phase… it's brutal. Jordan keeps telling me I need to trust the systems, trust the team, but everything's pulling me in a hundred different directions. I don't know if I'm cut out for this."

There was a beat of silence before Emily spoke again, her voice quieter this time. "You know, I've been thinking about that, too. Not about you, but about us. Our family."

Alex turned to her, surprised. "What do you mean?"

She leaned back, looking at him with a soft expression. "You're not the only one going through changes, Alex. When we started this whole journey, I thought it was just about the business. But as you've been growing as a leader, I've been growing, too. Watching you struggle, push, and fight for this—it's made me rethink my own role. Not just as your wife, but in our family."

Alex blinked, caught off guard by her words. "I didn't realize…"

"I know," she said, smiling slightly. "I didn't either, until recently. I've had to adjust, just like you. You're not the same man I married, and that's a good thing. We're both evolving. It's uncomfortable, but it's also exciting, isn't it?"

Her words hit him differently. This wasn't just about his personal growth anymore. Emily was evolving alongside him, adapting to his journey in ways he hadn't fully considered. He thought about the nights when he stayed late at the office, the weekends when his mind was somewhere else, consumed by the business. She had been there through all of it, quietly growing with him.

"I guess I never thought of it like that," Alex admitted, running a hand through his hair. "I've been so focused on the business that I didn't realize this was affecting you, too."

Emily laughed softly, shaking her head. "It's all connected, Alex. You've always been the one to dive headfirst into things, and I've been the one to make sure we don't sink. But lately, I've had to step up in different ways. And honestly? It's been hard. But I wouldn't trade it for anything. Watching you figure this out, even when it feels like the hardest thing in the world, it makes me proud of you."

Alex felt a warmth in his chest that he hadn't expected. For so long, he had seen the pressure as something he had to bear alone, but Emily's words opened his eyes to a new reality: they were in this together, evolving together.

"I don't know if I say it enough," Alex said quietly, his voice thick with emotion. "But I couldn't do this without you."

Emily smiled and leaned in, resting her head on his shoulder. "You don't have to. We've got this."

The Hard Calls: Confronting Tough Truths

Alex sat alone in his office, the warm afternoon light casting shadows across the room. His fingers hovered over the keyboard, but he couldn't focus. One thought gnawed at him, refusing to let go—the conversation he needed to have with Ethan.

For weeks, he'd told himself other things were more pressing. But the truth was clear: Ethan, his longtime friend and Head of Account Management, wasn't keeping up. The agency's growth was straining his abilities, and that was dragging them all down.

Alex sighed, leaning back in his chair. He knew that leadership wasn't just about landing clients or building systems. He had to face moments like these and he had to accept that not everyone who started with him could finish the race.

His phone buzzed, a text from Maya: "We're having issues with Ethan's accounts again. Clients are frustrated by the delays."

Alex's stomach tightened. This wasn't the first complaint. He'd hoped things would improve, that Ethan would find his footing. But deep down, he knew the truth. Avoiding it any longer would make things worse, not better.

With a heavy breath, Alex reached for his phone and dialed Jordan. After a few rings, his mentor picked up.

"Alex," Jordan's familiar voice greeted him, steady as always. "How's everything going?"

Alex hesitated, then let out a sigh. "Not great. I think... I think it's time for a hard conversation with Ethan. He's struggling, and I don't think he can keep up with the pace anymore."

There was a pause. "You've been feeling this way for a while, haven't you?" Jordan asked, his tone empathetic yet direct.

"Yeah," Alex admitted. "But it's not easy. He's been with me since the beginning."

"Alex," Jordan's voice softened, "I know it's difficult. But remember, this isn't about how long he's been there. It's about the agency's future—and his, too. Holding on to someone who isn't keeping up doesn't help either of you. The facts are clear: if he's holding back the growth, you're letting emotions dictate your decisions."

Alex exhaled slowly, the weight of Jordan's words settling in. "It's just hard to think about letting him go. He's been a big part of this."

Jordan's voice remained calm. "True leadership isn't about avoiding the hard calls. It's about making decisions based on what's right for the business, not just what feels comfortable. Ethan's a great person, but he's not meeting the demands of the role anymore. And that's not sustainable for anyone involved."

Alex stared at the wall, the truth of Jordan's words unavoidable. "You're right," he said quietly. "I have to face it."

"Good," Jordan replied. "This is the moment where you show yourself what kind of leader you are. Approach it with respect, but stay rooted in the facts. You owe that to the agency—and to him."

"Thanks, Jordan," Alex murmured, his voice thick with emotion. "I needed that."

After the call, Alex put down the phone, feeling the weight of his decision pressing on him. It was time.

The walk to Ethan's office felt longer than usual, every step amplifying the dread in his chest. Pausing outside the door, he could hear Ethan typing, unaware of what was coming.

He closed his eyes for a moment, centering himself. They'd built this agency side-by-side. But instinct and loyalty weren't enough anymore; they needed precision, discipline—things Ethan was struggling to deliver.

He opened the door.

Ethan looked up, his usual friendly smile in place. "Hey, man, what's up?"

"Got a minute?" Alex asked, forcing a smile he didn't feel.

"Of course," Ethan said, leaning back, unaware of the tension hanging between them.

Alex sat across from him, wrestling with his thoughts. For a second, he almost backed out, but he couldn't. If they didn't start making tough calls now, the agency wouldn't survive.

"Ethan," Alex started, his voice softer than usual, "we need to talk about how things are going."

Ethan's smile faded. He straightened, his expression turning wary. "Listen, I know things haven't been going great. But I think that if I just have a couple more weeks I think I can fix it all. I hope..."

Alex took a breath. "Look, the agency's growing, and it's pushing us in ways we didn't expect. The structure, the pace... it's all changing. And... I don't think it's working for you."

The words hung in the air, heavy and irreversible. Ethan blinked, his face paling as realization set in.

"I..." Ethan's voice cracked, and he swallowed, struggling to keep his composure. "I've tried, Alex. I'm working my ass off, but..." His voice trailed off, gaze dropping. "I'm not the right fit for this anymore, am I?"

Alex's throat tightened. This was harder than he'd imagined. "You've been with me since day one. But where we are now... we need something different. It's not about your past work. It's about what we need moving forward."

Ethan's hands clenched, knuckles white. "I know," he whispered. "I've felt it too."

The silence felt heavy. Ethan was more than a colleague; he was a friend. But the agency couldn't afford to be weighed down by the past.

"I'm sorry," Alex said quietly. "But I think it's time."

Ethan nodded, blinking rapidly. He stood up and moved to the window, staring out over the city. "If I'm being honest, I knew this day would come," he said, voice distant. "I just didn't think it would be so soon."

Alex joined him, struggling to find the right words. All he had was the truth.

"I couldn't have built this without you," Alex said, voice rough with emotion. "But I can't keep pretending it's working. It's not. And we both know it."

Ethan turned back, his jaw tight, but there was understanding in his eyes. "You're right." He let out a shaky breath. "I'll pack up my things. I'll be out by the end of the day."

Alex wanted to say something to soften the blow, but there was nothing left to say. He extended his hand.

"Thank you, Ethan. For everything."

Ethan hesitated, then shook his hand, his grip firm but trembling. "Take care, Alex," he said softly.

Ethan packed up his things and closed his laptop for the last time. Alex watched as Ethan left, the front door to the office clicking softly behind him. The silence that followed felt heavier than anything he'd ever known.

That evening, Alex sat on the porch, staring out at the dark abyss in the sky, but this time he felt more angst and sadness than in weeks past. The weight of the day pressed down on him, clinging to his chest.

Emily joined him, glass of wine in hand. She didn't say anything, sensing something was wrong.

"Rough day?" she asked quietly, taking a seat beside him.

Alex nodded, his gaze on the horizon. "I had to let Ethan go."

Emily's expression softened, and she placed her hand on his arm. "I'm sorry."

He let out a long breath, the tension easing slightly. "Yeah, me too."

They sat in silence, the cool evening air washing over them. Alex's mind raced with everything—the conversation with Ethan, the weight of the decision, the uncertainty of what came next.

"You did what you had to," Emily said gently. "Sometimes you have to make some really tough calls. And this was one of those times."

Alex nodded, her words settling over him. "I know," he whispered. "But it doesn't make it easier."

Emily squeezed his arm, giving him a small, sad smile. "No, it doesn't. But you're doing the right thing."

Leading with Purpose: New Responsibilities

The atmosphere in the office was tense yet focused. Ethan's departure had left a gap in Account Management, and the pressure on the team was tangible. Maya had taken on some of Ethan's client management responsibilities, Rebecca was balancing HR and stepping in where needed, and David was closely monitoring client revenue projections to ensure they didn't slip. Each team member was giving more, stretching their capacities, but they all felt the strain. Alex knew it couldn't continue like this for long.

Alex stood at the glass window of the conference room, the city skyline spread out before him. It had only been a couple weeks since Ethan was let go, but it already felt like much longer. They were on the hunt for a new Head of Accounts, someone who could bring expertise and a fresh perspective. The challenge, however, was finding someone with both the experience to handle a role of this scale and the right cultural fit—a person who could step seamlessly into their tight-knit team without disrupting the balance.

He turned back to the team, who were seated around the conference table, their faces drawn with exhaustion. Jordan had advised Alex to find someone from outside, someone who'd faced the very challenges they were working to overcome.

"Jordan's right," Maya said, her voice breaking through Alex's thoughts. "We need someone with a track record, someone who's done this before. I know it'll be hard to find, but we can't keep operating for long with the team stretched like this."

David nodded. "It's not just about filling this seat. This is about finding someone who aligns with our values and can lead this department effectively. Otherwise, we're going to face the same issues again down the road."

Alex nodded, feeling the weight of their words but wrestling with his own reservations. He was used to trusting people he personally connected with, people he felt he could work side by side with. But this wasn't about personal feelings anymore. It was about what the agency needed.

Over the next three weeks, they brought in a series of candidates. Most were solid, with years of experience in account management, but no one seemed to meet the high standards they had set in the job description. Then, they interviewed someone different—Lucas was a candidate with a strong record in client retention, proven leadership skills, and an impressive understanding of the systems they were trying to implement. His resume showed experience solving exactly the types of problems they were facing, but something about them didn't click for Alex on a personal level.

After the interview, Maya and Rebecca were visibly enthusiastic. "I think Lucas could be the one," Rebecca said, glancing at Alex with a hopeful smile.

David nodded in agreement. "He has everything we need—a track record, strategic thinking, technical ability, and clear examples of where he's improved client relationships in other agencies just like ours. He'll help us get the retention metrics we've been pushing for and can start building out that team. He's literally done everything we need to execute based on Jordan's checklist."

But Alex found himself holding back. There was something about Lucas that felt off to him, something intangible. He couldn't imagine working closely with him in the way he had with Ethan. Yet, the team's arguments were compelling; the facts aligned, even if his instincts didn't.

That evening, Alex called Jordan, hoping for clarity. He detailed his concerns, sharing his reservations about not feeling a personal connection with the candidate.

Jordan listened patiently before replying, "Alex, remember what we've been working on—facts, not feelings. Your team is excited about this candidate because he has the experience, skills, and track record to solve the exact problems you need to solve. If you're making this decision based on who you 'feel' connected to, you're not leading with your head."

"But what if he doesn't fit in with us, with our culture?" Alex countered, clinging to his instincts.

"Has he demonstrated the core values that matter most to you and your team? And I'm talking about specific examples where he solved problems the exact way that you would want him to solve your problems according to your core values." Jordan asked pointedly. "If he aligns there, then he's a good fit, even if he's different from what you're used to. You have to lead with a broader perspective, Alex. This is about what's best for the agency's growth."

The conversation hung heavy in Alex's mind as he walked into the office the next morning. He could see the signs of strain on his team—the tight expressions, the fatigue. This hire wasn't just about him. It was about finding someone who could take the weight off his team, someone who could bring the agency to the next level.

Alex gathered the team, ready to make the final decision. His chest tightened as he looked around the room at Maya, Rebecca, Sophia, and David, who all looked hopeful but cautious, waiting for his verdict.

"All right," Alex began, his tone steady but resolved. "We're going to move forward with Lucas."

A wave of relief washed over their faces. Maya gave him a small nod of approval. "You made the right call, Alex. I know this wasn't easy for you, but this is what we need."

He still didn't feel fully convinced, but he knew this was a moment to trust his team's judgment—and Jordan's advice. It wasn't about how he felt; it was about what was best for the agency.

The first few weeks were an adjustment. Lucas brought a fresh, disciplined approach that quickly alleviated the pressure on Maya and Rebecca. Processes that had once been bottlenecks were now running smoothly, and client feedback improved within the first month. The team's workload became more manageable, and there was a noticeable lift in morale.

Alex watched as Lucas interacted seamlessly with the rest of the team, winning them over with quiet competence and a clear understanding of the challenges they faced. For the first time in months, Alex found himself with breathing room. They were meeting deadlines, their retention metrics were climbing, and new clients were coming on board smoothly. The agency was running like a well-oiled machine.

And Alex felt something shift in himself. He realized he didn't need to feel a deep personal connection with every team member; what mattered was the value they brought and how well they fit with the agency's mission and values. This new approach, rooted in data and the team's needs rather than his own instincts, was the path forward.

The Final Stretch: Resilience Realized

On a crisp autumn morning, Alex arrived at the office, noting the clear, focused energy that now defined their agency. It had been over a year since he started implementing Jordan's advice and this latest Developing Level was months of struggle, doubt, and hard decisions but it had given way to confidence and stability. He felt a new sense of pride walking through the hallways, seeing his team working seamlessly within the systems they had painstakingly put in place. They had evolved into a true agency, where each person knew their role and the business operated like a well-oiled machine.

As Alex entered the conference room for their final Developing Stage review, Maya, Rebecca, Lucas, and Sophia were already gathered around the table, faces alight with a sense of accomplishment. He felt a surge of gratitude for the team and everything they'd achieved together.

"Morning, everyone," Alex greeted them, settling into his seat. "Today, we're here to wrap up Level 3. It's been a long journey, and I can see how far we've come. Let's check off those last items."

Maya opened the checklist on the projector, where only a few boxes remained unchecked beneath the words Developing Stage. Each remaining item was a testament to the progress they'd made in this intense year of growth.

"Alright," Maya began, a look of quiet pride on her face.

"We've been working through the last few items, and I'm pleased to report we're just about there. Rebecca, could you start us off with the HR update?"

Rebecca nodded, flipping through her notes. "We finalized the performance review system last month. Everyone on the team now has a clear roadmap for growth, and the new bonus structure is directly tied to these metrics. Morale has noticeably improved, with more employees taking initiative. It's not just about the high performers anymore—everyone has a role and understands what success looks like."

Alex smiled, feeling the gravity of the achievement. "It's amazing to see. We've gone from firefighting every day to building something sustainable."

He turned to Sophia, ready to hear the update on lead generation and sales closing. "Sophia, how's our lead generation and sales pipeline looking?"

Sophia nodded with a confident smile. "We've successfully diversified our lead generation sources. In addition to referrals, our new PPC ads and the trade shows we are exhibiting at have dramatically increased our lead flow, producing a steady influx of high-quality leads every single week. Our conversion rate on qualified leads is now at 25%, up from 15% just a few months ago since we created the new packages. The structured pipeline has also helped us close deals faster, so we're hitting all of our quotas."

Alex was really excited by this news. They'd worked hard

to reduce dependence on a single source of leads, and now they were reaping the rewards. "That's excellent work, Sophia. This diversification is exactly what we needed."

Next, he turned to Lucas, eager to hear about the progress in Account Management. "Lucas, what about client accounts? Any updates on retention?"

Lucas nodded, pulling up a slide on his laptop. "We've been able to bring our monthly churn rate down to under 3% for the first time in agency history. While we still experience occasional cancellations, we've established a proactive client journey that lets us spot potential issues before they become reasons to cancel. The retention system we put in place means we're never caught off guard, and the few clients we do lose are no longer hitting us as unexpected setbacks."

This was the dream they'd had when Jordan first pushed them toward developing a solid client journey. For the first time, their client base was growing consistently, supported by a retention plan that prevented the churn problems that had haunted them for years.

Later that week, Alex arrived at Jordan's office with the agency's most recent reports. They had completed every item on the checklist, and the numbers backed up their achievements: financial stability, retention gains, and a solid operational foundation. Jordan greeted Alex with a smile, motioning him toward the armchair opposite him.

Once seated, Alex began to walk Jordan through the final updates, explaining how each team member had contributed to the agency's success. He felt a rare surge of pride as he shared the details, seeing Jordan nodding in quiet approval.

Jordan folded his hands, listening carefully before he spoke. "Alex, I've been reviewing your numbers, and I have to say—it's impressive. You've not only met but exceeded every target we set at the beginning of Level 3."

He paused, looking at Alex with an expression of pride and satisfaction. "I want to talk through each of these areas, not just as an update but to reflect on what you've achieved—and to get a sense of where you go from here."

Alex nodded, sensing the significance of the moment.

"Let's start with finance," Jordan said, gesturing for Alex to elaborate.

Alex pulled up the most recent profit and loss statements on his laptop. "We hit a consistent gross margin of 70%, with net profits increasing every quarter. David's cost-cutting measures, combined with renegotiated contracts and our updated pricing model, have given us a stable, profitable foundation."

Jordan looked over the report, nodding. "You've turned the financial side around completely. A 70% gross margin is more than solid for an agency of your size. It means you're not just surviving—you're scaling, sustainably."

He leaned back in his chair. "Tell me, how does it feel to finally be in control financially?"

Alex thought for a moment before answering. "It feels... liberating. For the first time, I'm able to look at our finances without that pit of worry in my stomach. We're not just reacting to problems; we're planning for the future."

Jordan smiled. "Exactly. Financial stability doesn't just mean profit; it means freedom and growth."

"Now, tell me about your team's performance," Jordan continued. "How are they handling the systems you've put in place?"

Alex felt a surge of pride as he spoke. "The team has taken ownership of the processes we set up. Maya has the operations running with clockwork precision. Rebecca's HR systems have brought in a culture of accountability and reward, and the morale has never been higher. Each team member knows their role, and they're not waiting for instructions—they're driving the agency forward."

Jordan listened intently. "And how have you managed to step back and let them take the reins?"

Alex smiled, realizing the transformation within himself. "It was hard at first, but I began to see how empowering them made the agency stronger. Letting go has allowed me to focus on strategy rather than survival."

Jordan nodded approvingly. "That's the heart of the

Developing Stage—turning your team into leaders themselves. That's how an agency goes from founder-led to team-driven. And you've done it."

"Let's talk about client retention," Jordan prompted. "I saw that you've finally brought the churn rate down to below 3%."

Alex nodded, feeling another wave of pride as he explained. "Lucas transformed our client journey. With his experience and insights, we can now predict churn risks early and address them before clients consider leaving. This means we're keeping clients happy proactively, not just reacting when things go wrong. The few who do cancel are no longer a surprise, and our retention is stronger than ever."

Jordan leaned forward, impressed. "You've created a client journey that builds loyalty, not just satisfaction. This approach to retention is going to be a game-changer as you grow. And Lucas was the right call—facts over feelings worked here, didn't it?"

Alex smiled, acknowledging Jordan's guidance. "Yes, it did. Trusting the team's instincts and the data was the right move, even if it was uncomfortable at first. Lucas has been invaluable."

Jordan shifted topics again. "And what about sales? Sophia's diversified lead generation channel and closing rates—how's that going?"

Alex pulled up the reports Sophia had prepared, detailing their success. "Sophia's team achieved a 25% close rate and a steady lead flow. We're not reliant on one channel anymore, and the sales process has become smooth and reliable."

Jordan looked impressed, noting the significance. "Breaking out of that dependence on referrals means real growth potential. You're setting yourselves up for long-term success in client acquisition."

Jordan leaned back, folding his hands as he surveyed the accomplishments. "Alex, you've made it. Level 3 is completed. Your agency isn't just keeping its head above water. You guys are absolutely thriving. And you're seeing the financial and personal rewards of that transformation."

Alex felt a weight lift from his shoulders. "For so long, I felt like I was trying to hold everything together on my own. Now, I'm looking at an agency that runs without me having to steer every detail. It feels... freeing."

Jordan smiled, clearly proud of his mentee. "That's because you've built a business that outlasts you, one that doesn't need you in every decision but benefits from your vision. This is what it means to be a leader, Alex."

Alex took a deep breath, feeling the magnitude of everything they'd achieved. He looked at Jordan, filled with gratitude. "Thank you, Jordan. For everything. I couldn't have done this without your guidance."

Jordan clapped him on the shoulder. "You did this, Alex. I gave you the map, but you did the work. Now, it's time to move into the Professionalizing Stage and take this agency even further."

That evening, Alex took the long way home and drove through the city with a quiet sense of accomplishment. For the first time ever, the agency's success wasn't just a hope—it was a reality. The financial gains were substantial, their processes streamlined, and their client relationships solidified. They had grown in every way, and the future was wide open.

Emily joined him, leaning her head on his shoulder, a smile playing on her lips. "You're finally here, aren't you? Really here."

Alex nodded, feeling a quiet confidence settle over him. "We're ready, Emily. The agency's set up for growth, and I'm not holding us back anymore."

She smiled, wrapping her arms around him. "What's next?"

He looked out over the lights of the city, feeling excitement and peace. "Next, we take this agency to places I never imagined."

And What Might Your Marketing Agency Need for Growth?

If you're struggling to scale your agency and feel like you're running in circles, then you're not alone. Many agency owners face roadblocks when it comes to growth, often unsure where to start.

But what if you had a proven checklist that guides you through every critical step of scaling?
Inside this chapter, our character Alex talks about the very same Agency Growth Checklists that help him break free from the daily grind and build a business that practically runs itself.

Want to see the exact checklist that took him from chaos to control?
With actionable steps on everything from optimizing client fulfillment to refining team operations and boosting profit margins, this checklist can be yours.

Visit **www.factsnotfeelingsbook.com/checklists** and download all of our checklists for free and get the structure your agency needs to thrive.

Chapter 6: The Professional Shift
Level 4

Vision and Goals: Setting the Path with Jordan

The office was buzzing with anticipation as Alex gathered his team in the conference room, waiting for Jordan to arrive. Sunlight streamed through the floor-to-ceiling windows, casting a crisp clarity that mirrored Alex's hopes for the coming months. This was the Professionalizing phase—a chapter that, if executed right, would turn their agency into a scalable powerhouse. And today, for the first time, Jordan would meet the full leadership team, not just Alex. He'd asked to sit down with everyone to review the objectives of the checklist he'd meticulously designed.

As the clock struck ten, the door swung open, and Jordan walked in with the confidence and calm that Alex had come to respect. He was here not as an advisor to Alex alone but as a mentor to the entire agency.

"Good morning, everyone," Jordan began, glancing around the room. His gaze moved from Maya, who managed operations, to Sophia, heading up marketing and sales, then to David, who led finance, and finally to Lucas, the new Head of Account Management, to Rebecca the head of HR. Each of them sat at attention, aware of the gravity of this meeting.

Jordan leaned against the edge of the conference table, crossing his arms. "I've seen you all grow this agency from a scrappy, rudderless agency to what it is today—a thriving business with the potential to be one of the most respected names in the industry. You all have worked so hard to get here, and we're entering a phase where everything changes. We're not just keeping this business afloat now. We're going to make it even more scalable and even more profitable. And you're all going to lead the charge."

He let the words sink in, and Alex could feel the weight of Jordan's expectations pressing on each team member. Jordan opened a folder and slid a copy of the Professionalizing checklist across the table to each team member.

"This is your roadmap," he said, tapping the paper. "Nine months. That's the timeline we're working with, and every objective here is non-negotiable. Each of you has a critical role to play. You're going to master your departments, build systems that scale, and prepare this agency for something much bigger. Today, we're here to set the stage and map out exactly how you'll execute each objective."

Jordan's voice carried the kind of authority that commanded attention, and the team leaned in, ready for what came next.

Jordan flipped open his notebook, gaze steady as he looked to Maya first. "Maya, let's start with Operations. You'll be focusing on building middle managers in fulfillment and refining your SOPs so that when the time comes, the department can scale tenfold without breaking a sweat. SOPs must ensure consistent costs, high-quality deliverables, and a resilience plan—disaster recovery—to keep Operations running smoothly. Your department's success will depend on reporting systems that can handle up to ten times the clients you have now, with data that flows from fulfillment to client-facing teams, proving ROI at scale. I need you to find your replacements and groom them well." He paused, allowing Maya a moment to let the message hit.

Maya nodded, her eyes flashing with determination. Her posture straightened, already accepting the challenge. "I understand. I'll refine the SOPs to handle everything from day-to-day fulfillment to recovery, and I'll make sure the team is ready to support ten times the workload."

Jordan gave her a short nod of approval, then turned to Lucas, who seemed braced, the gears already turning. "Lucas, your focus will be on a watertight client experience. I want churn under 3% each month and zero 'red zone' issues. Your team should be proactively identifying problems and preventing churn before it starts—if a client is frustrated, you should know before they do. Your department must build a system for reporting that highlights issues early and gives clients the assurance they need. Part of that means a comprehensive client education program. You'll have to set the standards here and, like Maya, bring in managers who can carry this weight with you."

Lucas nodded, his expression intense. "I've already started mapping out some of the client reporting improvements. I'll make sure our team is managing proactively and that we're predicting churn patterns instead of reacting to them. I'll work closely with Rebecca to ensure we can expand seamlessly."

Jordan acknowledged him with a quick nod and shifted to Sophia, whose energy buzzed with an eagerness she could hardly contain. "Sophia, you're tackling marketing and sales. First, you need middle managers to help with daily oversight. You'll refine every SOP in place until we have absolute consistency. I want accurate tracking of customer acquisition costs down to the dollar, and I want you to drive new business beyond what we've relied on so far. That means developing lead generation systems and hosting events to elevate brand awareness. Sophia, I know you've been on the frontlines in sales, but this phase means expanding beyond individual achievements. You'll also need to build referral relationships to create an affiliate structure that supports growth."

Sophia nodded enthusiastically, a spark of excitement in her eyes. "I'll refine our lead gen and bring in new managers to handle the day-to-day. I know we can push our brand to a new level—events and referrals are where we'll make that happen."

"Good," Jordan said, satisfied but focused, as he turned to Rebecca, who was ready with notes. "Rebecca, with HR, you'll establish solid internal foundations. Your role is to shape the culture, to ensure this team stays as sharp and aligned as they are now. Build our company handbook with clear policies, structures, and career progression paths. And as for employee retention, work on incentive plans that keep people focused on growth. We also need to recruit and train middle managers. The company is getting bigger and if we have each manager overseeing more then 6 or 7 people we are going to start to see breakdowns. Finally, work on documentation—polished, complete, and prepared for Alex's potential exit."

Rebecca nodded, confidence steady. "I'm on it. Our culture is crucial, and we'll make sure every singe one of these tasks are checked-off the list."

"David," Jordan shifted his attention to the man responsible for the finances. "We are entering a really crucial time in the company. We must run our numbers with the same tenacity as a Fortune 500 company. You've heard this before...revenue is vanity, profit is SANITY. I need you to make sure that you are a pitbull with the budgets and make sure that we hit our profit goals. Can I count on you to do this?"

David beamed with pride and responded "You will not have to worry about a thing, sir. I will treat the company's finances as if they were my own."

Finally, Jordan turned his attention to Alex, an almost imperceptible smile on his face. "Alex, your task is to be in this business and out of it at the same time. You're developing your team to take over strategic decisions, creating scorecards and KPIs that show where the agency stands without you looking under every rock. Your focus is on high-level strategy and on preparing the agency for a potential exit, and that starts with documenting everything—every SOP, every metric, every milestone."

Jordan's words felt weighty, but Alex nodded, feeling the thrill of this new step. "I'm ready," he replied, voice steady but excited.

Jordan closed his notebook and addressed the team, his gaze sweeping across each of them. "In nine months, we'll revisit this room. I'll expect every department to have its systems, managers, and reporting in place. This agency will either become bulletproof or reveal its weaknesses." His eyes landed on Alex last. "And, Alex, you'll be ready to make a choice."

The room sat in silence for a beat, the collective energy buzzing with determination. Alex could sense his team's shared resolve, an unspoken promise to rise to the challenge.

The room cleared, and Jordan motioned for Alex to stay. The seasoned mentor studied him for a moment, the weight of their shared journey resting between them.

"You've come a long way, Alex," Jordan said, a hint of pride in his tone. "This is where your work as a leader steps up. If you're successful, you're building something that can last beyond you—and that's the real power."

Alex swallowed, feeling the enormity of what lay ahead. "And if I don't? If something slips?"

Jordan didn't blink. "Then you adjust. But remember, this isn't about saving the day anymore; it's about creating a self-sustaining entity. For you, this next stage is less about actions and more about restraint. Letting your team carry this, even if it doesn't go perfectly, will show you the agency's true resilience."

The conversation shifted, and Jordan leaned in, his voice lowering. "When these nine months are done, I'll be here again with your team, and if you've built what I think you're capable of, we'll start talking about your exit. That's not just freedom, Alex. It's the agency's legacy. Your legacy."

Alex felt a shiver of anticipation, the potential within his reach. As he walked out, he felt the weight, not of pressure, but of possibility. He had a team of people ready to make this happen, and for the first time, he knew they were just as committed to the vision as he was.

Operational Mastery: Building Strong Systems

Under the crisp, clear light of the morning, Alex's team gathered around the conference table, their faces a blend of determination and a hint of nerves. Sunlight streamed in, illuminating the room and casting a golden edge to the task ahead. This wasn't just another meeting; this was the start of the Professionalizing phase—a test of everything they'd built and of each leader's ability to carry the agency to its next evolution.

Maya sat beside Alex, flipping through the stack of carefully printed SOPs. She had reviewed each one a dozen times, scrutinizing them with the same rigor she applied to her team's deliverables. As she prepared to present her progress, she felt the weight of her dual responsibilities—managing day-to-day operations while crafting a structure that would eventually outgrow her.

"Maya, you've got this," Alex said quietly, sensing her tension. "Remember, it's a marathon, not a sprint."

Maya's lips tightened in a faint smile. "That's what I tell myself every morning. But letting go of these tasks feels like pulling teeth. I know I have to trust Priya and Tom to step up, but part of me is afraid we'll lose our edge if I'm not there to manage the details."

Her two chosen middle managers, Priya and Tom, were both ambitious and skilled, but they'd yet to handle oversight of this scale. Priya was sharp with operations, and Tom had a natural rapport with clients, but neither had led independently. Maya's challenge was more than just delegation; it was about allowing these two to find their footing—and trusting that they'd uphold the standards she'd set.

As the weeks unfolded, Maya faced moments of doubt. One evening, she found herself redoing an SOP that Priya had revised, unwilling to let go. But as she sat alone in the dim glow of her computer, Jordan's words echoed in her mind from an earlier check-in: "True growth happens when you let people learn through doing. Your job isn't to catch every mistake; it's to guide them in catching their own."

The next day, she called Priya and Tom into her office, handing them full control over the documentation process. "You've both got this," she told them, her voice carrying a new resolve. "It's time you own these projects."

Over the months, Maya watched them struggle, adapt, and ultimately flourish. Priya's confidence grew, her SOP revisions becoming crisp and precise. Tom, who initially struggled with balancing client expectations, started using the reporting systems to anticipate client needs and prevent potential issues. By the end of the quarter, Maya knew she had made the right choice. Her department was stronger for it—and so was she.

<hr />

Sophia, seated across the table, exuded her usual calm determination. She was a natural at inspiring confidence in the team, but the Professionalizing checklist demanded more than just charisma. To get the agency where it needed to be, they couldn't rely solely on word-of-mouth and referrals. They needed a steady, measurable flow of new leads, and Sophia would have to break new ground to get there.

Her first target was tracking CAC with razor-sharp accuracy. For weeks, she immersed herself in analytics, fine-tuning every metric and working with David to ensure their costs and returns were as clear as possible. "We're investing in this growth," she reminded Alex and David in their next budget meeting. "But without hard data, we're flying blind."

But it wasn't just about numbers. Sophia had big ideas for brand visibility. She organized two major industry events, strategically positioning The agency as a thought leader in the digital marketing space. The first event was nerve-wracking; turnout was lower than expected, and her team voiced concerns about the investment.

That night, Sophia met with Jordan. "What if I'm overreaching?" she admitted, frustration slipping through her usually composed exterior. "Maybe we should just stick with what we know."

Jordan's eyes softened. "Sophia, you're not just building leads—you're building reputation, recognition, and momentum. The results won't be instant, but they will compound. You've got to see beyond the first event."

Taking his words to heart, she pushed forward with the second event, implementing feedback from her team and reworking their outreach strategy. This time, they saw double the attendance and an influx of new leads. The experience was a turning point, not only for the agency but for Sophia herself. She had proven that sometimes, growth required a leap beyond the familiar.

Lucas, the recently appointed Head of Accounts, had made strides with client retention, reducing churn to below 3%—a benchmark they'd struggled to hit consistently. With that target met, he knew the next challenge would be solidifying the systems that ensured clients stayed, not just for a cycle or two, but over the long haul. His focus shifted to building proactive trust through deeper engagement and developing a robust, scalable support framework that anticipated client needs.

Lucas initiated new early-warning indicators that flagged potential issues before they grew. He and his team implemented systematic client health checks, from weekly campaign performance reviews to routine follow-ups, all aimed at demonstrating the agency's value at every step. Additionally, Lucas launched an educational program that helped clients understand the full range of services available to them and how to leverage these for optimal results. His goal was simple: to make sure clients felt empowered and consistently saw the value in their partnership.

Early one morning, Lucas called Alex in. "We're sustaining churn below 3%," Lucas said, showing the latest client metrics on his screen. "The check-ins are working, and client confidence is rising. They're seeing our value without us having to scramble on the backend."

Alex nodded, visibly impressed. "That's the missing piece, Lucas. We're finally getting ahead instead of playing catch-up."

With Lucas's strategic foresight, client renewals reached new highs, and for the first time, the agency was stepping into a phase of sustained, proactive client care—no longer just reacting to challenges but anticipating and meeting them head-on.

David, known for his meticulousness, was a steadying force within the agency. Yet the Professionalizing checklist tasked him with something outside his comfort zone: building a positive, engaged company culture. He was skilled at creating incentive structures and overseeing financial metrics, but he had to delve deeper into what motivated his team on a personal level.

To tackle employee retention, David introduced performance-based incentives tied to both individual and team goals. The idea was simple: reward people for contributing to the agency's growth, while fostering camaraderie and shared purpose. At first, there was resistance. Some team members questioned the fairness of performance-based bonuses, worried about increased pressure and competition.

David, sensing the tension, reached out to each department head, organizing roundtable discussions to hear their concerns. "This isn't just about numbers," he explained in one meeting. "It's about making everyone feel they're valued—not just for the work they do, but for being part of the agency."

Over time, he worked with Rebecca to refine the incentive plan, incorporating feedback to create a more balanced system that rewarded collaboration as much as individual achievement. The two also collaborated to roll out a comprehensive employee handbook and structured onboarding processes, morale started to improve, and turnover rates stabilized. By quarter's end, David's work had cultivated a workplace culture that valued growth and retention in equal measure.

⸺ ◆ ⸺

For Alex, the Professionalizing stage was as much about personal growth as it was about operational goals. As each team leader navigated their checklist items, Alex worked closely with Jordan to develop accountability structures and CEO dashboards that allowed him to lead with vision rather than reaction.

One late afternoon, Jordan pulled Alex aside after a team meeting. "You're starting to see the pieces come together," Jordan observed, his tone approving. "How does it feel?"

Alex hesitated, realizing that for the first time, he felt more relief than worry. "It's strange," he admitted. "Letting go of control was terrifying, but now... now I see they're doing things even better than I would."

Jordan nodded, satisfied. "You're becoming the leader this agency needs. And if you stay on track, we'll have you ready for the final stage—Exiting."

As they continued their work, Alex felt the weight of Jordan's words. This wasn't just a step forward for the agency; it was a step toward a vision he had once doubted he could achieve. And now, with his team by his side, he was beginning to believe.

Becoming the CEO: Alex's Transformation

As the days settled into a rhythm, Alex felt the enormity of what was ahead. Jordan had drawn up the Professionalizing checklist like a roadmap, but Alex was beginning to realize that this phase was about far more than tasks and checklists. For him, it was about becoming the kind of leader who could let go and let the agency thrive independently. The journey toward hiring a COO would be central to this transformation, and as Jordan subtly reminded him, the right person wouldn't just improve the day-to-day running of the business. This person would lead the company into its next chapter.

Hiring a COO was an incredibly important undertaking and Jordan quickly laid things out directly. "Alex," he said, leaning forward with a quiet intensity that silenced any debate, "if you want to build an agency you can sell, you have to separate yourself from the daily operations of the entire company. I'm talking about you ONLY making strategic, high-level decisions. That means you need someone capable and, more importantly, trusted to run the day-to-day."

Alex nodded but felt an immediate twinge of hesitation. "I understand," Alex began, choosing his words carefully, "but this isn't just about letting someone else step in. It's about finding someone I can genuinely trust with what I've built."

Jordan nodded knowingly. "That's exactly right. And that's why the process will take time, patience, and an unshakeable focus on finding the right fit. I'm talking at least 4 months, maybe as long as 9 months of grinding the candidates down until we are 1000% sure it's the perfect person." With that, they dove into creating a blueprint for the ideal candidate, outlining a COO with a proven record in agency operations, a knack for leadership, and a sharp sense for numbers. The process would demand time and diligence.

Over the following months, Alex and Jordan began a series of meetings and interviews. They vetted candidates from top firms, many of whom had stellar qualifications on paper. Yet, as each meeting ended, Alex would walk away with lingering doubts. Either they lacked the presence that he imagined for his COO or, worse, they came across as too distant, too corporate, lacking the culture he'd worked so hard to build. But Jordan remained unfazed, reminding him time and again that they were searching for someone extraordinary, someone who could take the agency from "good" to "exceptional."

As the search continued, Alex started to focus on the larger picture of his own role. Under Jordan's guidance, the CEO dashboard, complete with real-time metrics on agency health, department scorecards, and predictive tools for client retention and acquisition started to improve and evolve. At first, stepping back from running the agency to focus purely on these metrics felt like giving up the thrill of being "in the action." Yet as he watched the metrics improve week over week, he saw that he wasn't really needed anymore in all the departments or any of them for that matter.

Then, after nearly two months of searching, Jordan introduced Alex to Michelle. Her resume was impeccable: ten years of experience managing agencies of similar size, a data-driven approach to team leadership, and a calm, assertive presence that balanced empathy with accountability. But it was her track record with past clients that stood out most—she'd built reporting structures that improved client ROI, successfully reduced churn across teams, and scaled agencies into new markets with sustainable growth.

In their first meeting, Alex immediately felt both intrigued and a bit intimidated. Michelle was clear, direct, and not particularly emotive. She had a logical, strategic air that almost seemed too polished. Alex had always gravitated toward leaders who connected emotionally, who could share a drink or a laugh. But here was Michelle, focused, determined, and calculating. The meeting wrapped up quickly, and Michelle left him with a firm handshake, her words still lingering in the room.

After she left, Jordan turned to him with a raised eyebrow. "Well?"

Alex hesitated. "She's... different," he admitted. "I'm not sure she fits in the way I'm used to."

Jordan nodded thoughtfully. "I expected you'd feel that way. Just how you felt about Lucas. Remember that? But here's the reality, Alex. I want you to consider the facts again—not just your gut feelings. She's solved the exact challenges you're facing now. And she aligns with the agency's core values: accountability, transparency, and results."

Jordan's words lingered, stirring a shift in Alex's perspective. He realized this was another moment to apply what Jordan had instilled from the beginning: facts, not feelings. He was holding onto a vision of leadership that was no longer suited for the agency's future. He had to break that mold and recognize what his agency truly needed, not what felt comfortable.

After several more interviews and tests, Jordan and Alex were very close to making Michelle an offer. However, there was one final test for Michelle- meeting the rest of the leadership team. Each leadership team member would meet with Michelle during the course of a week. It was more like a weeklong trial than an interview. Alex watched as Michelle connected with them seamlessly. Maya, normally reserved around newcomers, seemed engaged, even inspired by Michelle's methods. Lucas, typically protective over the client accounts, shared his vision openly, and Michelle's suggestions only seemed to enhance it. Watching her navigate the conversation, asking direct questions while also encouraging each leader's input, Alex saw a side of her he hadn't expected—a capability to empower others, even with her meticulous focus.

The team's response was unanimous. Maya, Lucas, Sophia, and David each expressed their confidence in Michelle, with Maya even voicing a hope that Michelle's presence could bring new stability to operations. "We need someone who has already crossed these bridges," she said, glancing at Alex. "Someone who can see beyond the details and create cohesion at scale."

Jordan added his final assessment after everyone had left. "She's exactly what you need, Alex. Your team believes in her, and they're right. Now it's your move."

On the final day of their weeklong interview, Alex called Michelle into his office to make the offer. He explained the salary, a substantial $350,000 base plus bonuses—a sum that took his breath away for a moment as he said it. Michelle, however, remained steady, absorbing the details without a flicker of hesitation. She accepted the offer with her characteristic composure, offering a handshake that felt like the beginning of something momentous.

The transition was carefully structured. Over the next few months, Alex worked closely with Michelle, gradually handing over operational responsibilities. She moved quickly, implementing a series of changes that surprised even Alex. She overhauled client reporting, instituted team performance reviews tied to individual and agency-wide goals, and initiated weekly check-ins with each department, building accountability and refining processes along the way. Her impact was undeniable, and for the first time, Alex found himself with something he hadn't truly experienced before: freedom.

With the COO role filled, Alex could finally focus on high-level strategy and preparing for the future. He and Jordan began mapping out the agency's growth trajectory, identifying potential markets for expansion, and shaping the narrative of an agency that could one day attract buyers. As he immersed himself in this strategic vision, he felt a profound shift in his identity. He wasn't merely a business owner anymore. He was a CEO and the architect of something that had the power to scale beyond him or any other individual.

In one of their final meetings before Michelle officially took over, Alex shared a moment with Jordan that was both humbling and exhilarating. "I didn't realize how much I was holding on to," he admitted. "This whole time, I thought I had to be everywhere, in every meeting, solving every problem."

Jordan smiled, nodding. "Growth isn't about doing more, Alex. It's about doing what matters and trusting others to handle the rest."

It was a simple truth, but one that had taken Alex years to understand. As he looked around the agency, he felt a strange mixture of nostalgia and pride. This was no longer just his creation; it was a living, breathing entity with a future all its own.

The Final Countdown: 60 Days to Go

The team gathered in the conference room, knowing they were in the final stretch. There was an unspoken energy in the air. For the last few months, they had all pushed themselves to their limits. But with Michelle's arrival, their collective momentum felt unstoppable.

Alex, now removed from all the day-to-day interaction with the team, looked around the room, noticing how each person seemed focused yet relaxed in a way he hadn't seen before. They were confident, more assured than ever, and it hit him how much each had grown into their roles.

Michelle, standing at the head of the table, flipped through her notes with a focused expression. "Alright, everyone," she began, her tone calm but laced with excitement. "We're officially in the final sixty days. Today, we'll tackle the last key items on the checklist. Let's keep things tight and efficient."

Sophia leaned forward, a spark in her eyes. "Lead generation has been hitting all targets, and CAC tracking is yielding solid numbers. We're acquiring customers for under $1500 which is 25% less than what we had budgeted for. And now we're seeing a steady flow of new clients without exhausting our resources," she reported, then looked to Michelle, as if seeking confirmation.

Michelle nodded, a slight smile breaking her usual focus. "Sophia, you've really created a sustainable process here. I've worked with many heads of marketing, and I have to say, your eye for precision in the numbers is rare."

Sophia grinned, a bit surprised at the direct praise. "Thanks, Michelle. It's been a team effort."

Alex took a mental note, realizing how Michelle's encouragement was subtly shifting team dynamics, adding confidence and trust. He glanced at Jordan, who sat observing at the far end of the table, arms crossed with a look of quiet satisfaction.

"David," Michelle continued, "Where are we on finance and employee retention?"

David pulled up a detailed graph on the screen. "Revenue and profits are at a record high. We're now at a $5M run rate which is more than 2.5x better than our best month last year. Then with Rebecca's new middle management hires, the overall refined hiring processes, and the improvement in documentation we have also significantly lowered our employee turnover. Now that we have the right people in the right seats we're currently meeting our client retention targets," he explained. He then turned to Alex, adding, "I think it's fair to say we're no longer firefighting on this front."Alex felt a swell of pride as he looked at David, who only a year ago had been navigating major finance hurdles. "Solid work, David," he said, meaning it deeply.

Then Michelle turned to Lucas. "Lucas, client experience is a game-changer. What's the status on churn and client satisfaction?"

Lucas glanced at his notes before speaking, his tone proud yet humble. "Our churn rate is at an all-time low of 2.8%," he said, "and it's not just about retention. We've implemented early warning systems and quarterly client education sessions. Clients feel valued, and they're recognizing the ROI without us needing to scramble to prove it. We're getting proactive rather than reactive."

Alex gave an approving nod. "That's exactly what we aimed for when we started all of this. Nicely done, Lucas."

After Alex gave an approving nod. "That's exactly what we aimed for when we started all of this. Nicely done, Lucas."

Maya, always consistent in hitting her goals, delivered on every single KPI and metric that she set out to accomplish. Everyone on the team knew that they could count on her and she raised everyone's performance because she set the bar so high.

Jordan leaned forward for the first time, his voice breaking the room's focus. "You've all done impressive work," he said, his gaze steady and appreciative. "You're showing the resilience and strength of a true professional agency. But remember," he added with a slight smile, "this is just the beginning."

Alex felt his heartbeat quicken. He'd come to see Jordan's praise as rare, almost sacred, and he knew his mentor didn't offer it lightly.

As the team wrapped up their reports, Alex caught Jordan's eye, and Jordan gave a slight nod, signaling it was time for their private conversation. Jordan followed Alex into his office, Alex's anticipation growing.

Jordan shut the door and gestured to the chair opposite

his desk. "Alex," he began, his tone softer now, more reflective. "Watching you and your team evolve over the past year has been a privilege. What you've built—it's something remarkable."

Alex replied confidently,"Thank you, Jordan. It hasn't been easy, but with everyone in their roles... I can actually see this agency working without me."

Jordan nodded, a glint of satisfaction in his eyes. "That's what a saleable business looks like, Alex. And I have to say, I've been thinking more about that exit plan we discussed."

Alex felt his stomach twist slightly with excitement. "Really? So you think we're close?"

Jordan leaned back, his gaze intense but warm. "Closer than ever."

"Michelle's been a game-changer," Alex admitted, running a hand through his hair. "I can see the team moving faster and making decisions without running everything by me. They respect her. Honestly, she's taken things further than I expected."

Jordan leaned in. "That's the mark of a strong COO, Alex. She's the buffer, the glue that keeps things moving without you needing to jump in. And the best part? She compliments your leadership style."

Alex nodded, a sense of relief settling in. "I'm ready to

trust her fully. I just need to learn to... let go."

Jordan chuckled, clapping him on the shoulder. "You've already started, Alex. Now, let's keep it going. With these final sixty days, you're not just preparing the agency for an exit. You're cementing a legacy. When this checklist is complete, you'll have something far more than a business—you'll have a self-sustaining, profitable machine."

Alex exhaled slowly, letting the weight of Jordan's words settle in. "I'm ready," he said, voice steady.

As they wrapped up, a knock came at the door. Michelle poked her head in, a confident smile on her face. "Sorry to interrupt," she said, "but the team's asking if we can do a final walkthrough of the checklist. They're eager to wrap things up."

Jordan raised an eyebrow at Alex, a silent nod of approval. Alex smiled, a genuine and unguarded expression of pride. "Let's do it," he said.

Together, they walked back to the team, ready to finish what they'd started, fully aware that this was only the beginning of something even bigger.

Reflecting on Growth: A Meeting with Jordan

For the past nine months, the leadership team and Alex had been climbing, executing on each part of the Professionalizing checklist with a precision that had transformed the agency. This was the day to celebrate their achievements and, as they all knew, to mark a turning point with Jordan.

Jordan began, his gaze moving around the room to meet each leader's eyes. "You've done something remarkable," he said, his tone both formal and proud. "Nine months ago, we laid out one of the most ambitious goals this agency has ever faced. Every one of you—Maya, Lucas, Sophia, David, Rebecca and Michelle—has taken ownership of that vision and brought it to life."

Maya nodded, a small smile breaking through her usual calm demeanor. She had led her team in implementing scalable systems and SOPs that had enabled them to handle a client load ten times what they could before. As Jordan turned to her, he continued, "Maya, you tackled some of the most challenging operational shifts. Thanks to you and your team, every project is executed with consistency and efficiency. The disaster recovery plan, the client reporting system, the scalable fulfillment—all of it has given this agency the foundation it needs to grow without limits."

The room broke into a round of applause, and Maya's smile widened, her confidence evident.

Jordan's gaze shifted to Sophia next. "Sophia, your efforts have paid off in ways that go beyond the numbers. Your lead generation systems and CAC tracking have brought in new clients at an unprecedented rate. Not only that, but the brand visibility you've created has made this agency a name people know and trust. I think we can all agree that we're no longer relying on luck or word of mouth."

Sophia gave a slight nod, a sense of accomplishment settling over her as she exchanged glances with Alex. The strategies she had put in place had paid off, propelling them toward a new level of brand authority.

Then Jordan turned to Lucas, who had redefined client retention. "Lucas, you've brought client churn to levels we once only dreamed of. And now, you're able to anticipate and address client concerns before they even think to raise them. The trust you've built between the agency and our clients is something few agencies ever achieve, and it's an invaluable asset moving forward."

Lucas, always calm and collected, gave a modest nod, his confidence showing in his posture. His department had become a machine of proactive care and client success, each interaction designed to reinforce the agency's commitment to its clients.

Finally, Jordan addressed David, the quiet but steadfast financial strategist. "David, this agency's profitability has become a testament to your work. The margins are healthier than they have ever been. Every financial metric has moved in the right direction, and that is no small feat in a company that has grown as fast as this one."

Then, Jordan turned to Michelle. "Michelle, in the short time you've been here, you've proven to be exactly the leader this team needed. The systems you've streamlined, the middle managers you've empowered, and the efficiency you've brought to every department have enabled Alex to step back. Your work has given the agency a new rhythm, one that doesn't depend on any single person but instead functions as a unified whole."

Michelle inclined her head, her expression one of quiet determination. She had come into a whirlwind and transformed it into something predictable, scalable, and dependable.

Jordan paused, letting his words sink in, the weight of the accomplishment resting on everyone's shoulders. "What you've accomplished here isn't just a list of completed tasks. You've transformed this agency into a business with systems, structure, and a strength that will allow it to thrive and expand without relying on any one person, even Alex." He looked directly at Alex, who stood off to the side, taking in the pride and admiration from his team.

Turning back to the group, Jordan's tone grew warm. "Today, I want you all to know that you didn't just meet the Professionalizing checklist—you redefined what it means to run a professional agency. And you should take a moment to acknowledge that achievement, because it's no small thing."

There was a beat of silence before Maya finally spoke up, her voice steady. "Thank you, Jordan. And thank you, Alex. I think I speak for all of us when I say that none of us could have done this without each other."

The room felt lighter, the weight of the past nine months lifting, replaced by a shared sense of accomplishment. Each leader knew they had reached a peak, and from that vantage point, the horizon looked clearer than ever.

With that, he stepped back, giving the team a nod of respect and a moment to take in the accomplishment. They had reached a new level, not only in their agency's journey but in their own professional evolution. And, as Jordan had said, it was only the beginning of what was next.

In the quiet of his office, Alex watched the city lights blink to life, stretching out into the horizon. The office had emptied for the night, the energy from earlier replaced by a hushed calm, like a deep exhale. Jordan had stayed behind for a private meeting, one that Alex knew would mark the start of something entirely new.

Jordan stepped in, closing the door softly behind him. He took a seat across from Alex, his expression contemplative, yet with a hint of a smile. "Alex," he began, "you've come so far from where we started."

Alex nodded, reflecting on the past months. "It's hard to believe. When I think about the fires we were putting out daily, where the team was back then... it feels like a different world."

"And you built that world, piece by piece," Jordan said. "But here's the thing—there's an even bigger shift coming. You're ready to take this agency beyond you, beyond any one person." He leaned forward, his gaze unwavering. "That's why I've been having conversations behind the scenes, Alex. I already have a strategic buyer lined up, someone interested in taking this to the next level."

Alex's eyes widened. "You... what? You've already found a buyer?"

Jordan smiled. "I knew it would be the right move, especially with how quickly you and the team were moving through the Professionalizing stage. They're a strategic buyer, someone who's looking for a high-functioning agency that runs like clockwork—something exactly like what you've built."

Alex felt a jolt of excitement mixed with a nervous edge. "And they're serious?"

"Very serious," Jordan replied, his tone steady. "They're the type of buyer who understands value beyond numbers on a balance sheet. They're looking at your systems, your leadership team, and the client relationships you've built. I've worked with buyers like them before, and I'm confident they'll see the full potential of what you've built here."

Alex leaned back, trying to process it all. "So... what's next? I mean, I didn't think this would happen so soon."

Jordan gave him a knowing look. "You've been ready, Alex. You've just been so entrenched in the work, you haven't seen the big picture shift in front of you." He paused, then continued, "I want to help you negotiate the deal. I'll be by your side through the entire process. This is the culmination of everything you've done—and what we set out to do from the very start."

A sense of accomplishment flooded Alex, tempered by the gravity of what lay ahead. He had always known this moment would come, but now that it was here, it felt surreal.

"And, Alex," Jordan added, his tone turning serious, "this next phase will be like nothing you've experienced. The Exiting stage isn't just about closing a deal. It's a transformation that will change your life. When this sale goes through, it's going to be a new world for you. Financially, professionally—even personally."

Alex took a breath, grounding himself in the reality of what Jordan was saying. "I'm ready," he said, the conviction clear in his voice. "I've worked too hard to turn back now."

Jordan nodded, his approval evident. "Good. Over the next few months, there will be due diligence, negotiations, and plenty of paperwork. But every late night, every rough conversation with investors, it'll all lead to something bigger."

The room settled into a deep silence, both men absorbing the weight of what was coming. Alex looked at Jordan, realizing that this wasn't just a business transaction; it was the culmination of a journey he hadn't known he was capable of completing. And standing on the edge of that precipice, he felt the pull of both accomplishment and anticipation.

As Jordan rose to leave, he extended his hand, meeting Alex's gaze. "You did it, Alex. Now let's make sure the world knows it."

Alex shook Jordan's hand and then hugged him with excitement. And just then he thought to himself, "I need to call Emily!".

And What Might Your Marketing Agency Need for Growth?

If you're struggling to scale your agency and feel like you're running in circles, then you're not alone. Many agency owners face roadblocks when it comes to growth, often unsure where to start.

But what if you had a proven checklist that guides you through every critical step of scaling?
Inside this chapter, our character Alex talks about the very same Agency Growth Checklists that help him break free from the daily grind and build a business that practically runs itself.

Want to see the exact checklist that took him from chaos to control?
With actionable steps on everything from optimizing client fulfillment to refining team operations and boosting profit margins, this checklist can be yours.

Visit **www.factsnotfeelingsbook.com/checklists** and download all of our checklists for free and get the structure your agency needs to thrive.

Chapter 7: The Crossroads
Level 5

A New Chapter: Meeting the Buyer

The boardroom gleamed with the polished surfaces and corporate precision that came with success. Alex could feel the intensity as he waited with Jordan for the arrival of the representatives from ClearEdge CRM, a major player in the software industry. ClearEdge, a publicly traded giant, had built a name by delivering customer relationship solutions that empowered companies to drive and nurture customer loyalty. Now, they were looking to take a step further by acquiring a strong marketing agency—one capable of elevating ClearEdge's service and reach.

Alex adjusted his tie, trying to stay calm, though he couldn't shake the slight tremor of excitement and disbelief. Just a year and a half ago, he was laser-focused on getting his agency's internal structure rock-solid; now, he was on the verge of potentially selling the business for over $18 million.

Jordan, seated next to him, leaned over and whispered. "I know this feels big—and it is. But remember, they're here because of what you and your team have built. You've put in the work, Alex. Don't lose sight of that."

The words calmed Alex, grounding him just as the doors opened, and a group of ClearEdge executives entered. At the forefront was Sean Hastings, ClearEdge's VP of Strategic Development, a man in his early fifties with sharp eyes and a confident, measured stride.

"Alex, Jordan," Sean greeted them, extending a firm handshake. "We're thrilled to finally meet face-to-face. ClearEdge has been looking forward to this conversation for a long time."

Introductions were brief, and the executives quickly got down to business. Alex noted their energy—direct and purposeful, yet leaving room for collaboration. It was exactly the tone Jordan had prepared him for: a team who knew what they wanted but respected the process.

"Alex," Sean began, glancing around the room. "Your agency stands out in today's market. We've seen the data, we know your reputation, and, frankly, we're impressed with the level of systems and leadership you've built. You've created a strong foundation that we believe can truly integrate and grow within our organization."

Alex felt a spark of pride. This wasn't a compliment to be taken lightly from a firm like ClearEdge.

Sean continued, his tone crisp, "Let's get into the terms we're offering. We've run the numbers, and based on the impressive EBITDA and the market positioning you've achieved, we're offering 12x EBITDA as our valuation. Given your current metrics, that puts us at a valuation of roughly $18 million. It's an ambitious investment on our part, and we're confident you'll see why it's worth it."

Alex glanced at Jordan, who gave him a subtle nod, a silent reminder to take in the information calmly. The number hung in the air, its weight sinking in as Alex digested it.

"Of course, there are steps we'll need to take to move forward," Sean added, bringing Alex back to the present. "We'll need a period of due diligence. Our team will review financials, processes, client contracts, compliance, intellectual property—standard protocol for any acquisition of this scale. It'll be intensive, but our team will be there every step of the way to make it smooth."

Jordan leaned in, his voice calm but assertive. "Alex and I anticipated a thorough review, Sean. We're prepared to provide the documentation you need. We'll coordinate closely with your team to address any requirements that come up."

Sean smiled, pleased. "Perfect. I think you'll find our team thorough but fair. We understand the importance of respecting what's been built here."

With the initial terms and approach established, the meeting shifted to logistics. ClearEdge's financial advisors laid out a tentative schedule, and Alex felt a rush of excitement tempered by a sense of responsibility as the scope of the process came into focus. Each aspect of his business—from client retention rates to employee retention and even cultural fit—would be scrutinized.

As the meeting wrapped, Sean rose and extended a hand to Alex. "This is just the beginning, but I have a strong feeling about this. We're excited to see where this takes us."

Jordan waited until the room had cleared before turning to Alex. "It's real now, isn't it?"

Alex let out a slow breath, his face a mixture of excitement and gravity. "It really is. I mean, 18 million, Jordan. That's... beyond anything I imagined at the start."

Jordan smiled knowingly. "That's because you were focused on doing the work. And now, with their offer on the table, it's time for the next level of work. Due diligence is where you'll feel every detail of what you've built."

As they left the office, Alex's mind whirled. Every corner of his business, the details he'd painstakingly set in place, would soon be dissected and evaluated. The process ahead promised to be challenging, but he was ready. Jordan's steady presence beside him felt like an anchor, keeping his perspective sharp as he took the first real steps toward a potential sale that could change his life forever.

Adapting Under Pressure: Adjusting to Change

Late one evening, Alex sat alone in his dimly lit office, the quiet hum of his computer screen casting a cool glow over the towering stacks of paperwork. He was utterly absorbed, reviewing every line item from client contracts to intellectual property clauses, each small piece of his agency laid bare for ClearEdge CRM's scrutiny. He was so immersed he almost didn't notice when Emily entered, a familiar warmth breaking the sterile, high-stakes atmosphere.

"Thought you might need something to remind you there's life beyond these four walls," she said, setting a warm meal beside his hand.

He looked up, his eyes tired but grateful, and for a moment, the weight lifted. "You always know," he murmured, squeezing her hand. "Feels like they're going through everything I've ever built with a microscope, like it's all on trial."

Emily sat beside him, her presence steadying as he navigated the flood of requests ClearEdge had sent. "This is everything you've worked for, Alex," she said quietly. "It makes sense they'd want to understand it all—but remember, you've done the work. You're not alone in this." Her words lingered as she placed her hand on his shoulder, a touch of calm amid the whirlwind of demands.

Over the following weeks, Alex dove deeper into the grueling due diligence process. He worked through late nights and long weekends, every detail scrutinized by ClearEdge's team, the sheer weight of the process sometimes eclipsing his excitement about the deal.

One night, he turned to Jordan, exhausted and feeling vulnerable. "They're looking at every detail," he confessed. "Every minor flaw, every gray area. I'm starting to feel like they'll find something, some crack, that could unravel it all."

Jordan leaned in, his expression focused but understanding. "That's the reality of due diligence, Alex. You're on the threshold of something massive, and they're making sure it's solid. But that's why I'm here. We're going to make this airtight."

Jordan identified areas that needed tightening: client contracts required small but critical adjustments, reporting systems needed refinement to meet data requirements, and intellectual property rights demanded clearer documentation. Each suggested change felt like a hit to Alex's pride—proof that, even after years of work, his agency still had hidden vulnerabilities.

"Without you here to guide me through this, I'd be totally lost," Alex admitted one night as they finalized another round of adjustments.

"That's what this is all about," Jordan replied, his voice steady. "Think of this as proof of just how valuable this agency is—not just because it's polished, but because you're willing to put in the work to make it last. ClearEdge sees that, and that's why they're here."

As Alex worked through ClearEdge's latest list of demands, the enormity of what he'd created settled over him. The process forced him to face every choice he'd made over the years. Yet, each adjustment also sharpened his vision, bringing him closer to a moment he both anticipated and feared. The stakes had never been higher, and as he looked around his office, he felt the weight of both his ambition and the legacy he was shaping—step by step, piece by piece.

The Legacy Question: Preparing for the Future

Alex sat in his office for what felt like the 50th night in a row, letting the silence settle around him, amplifying every doubt and hesitation. He was on the verge of making the biggest decision of his life, yet somehow it felt like an enormous loss, like letting go of a piece of himself. The office, once buzzing with ideas and endless possibilities, now held an unfamiliar quiet that mirrored the uncertainty in his chest.

Jordan's words had echoed in his mind all day. "Selling isn't just a transaction, Alex. It's a hand-off, a legacy." He had challenged Alex to consider the impact on everyone involved—especially the team that had put their trust and future in him. He knew he was selling more than a business; he was handing over the people and culture that had defined his career and, in many ways, his identity.

Later that evening, Alex found himself at home, leaning against the kitchen counter, lost in thought. Emily noticed the tension in his shoulders and quietly stepped beside him, gently placing her hand on his arm. "You've been in your head all day. Talk to me. This decision... it's eating you up, isn't it?"

Alex let out a long breath, the weight of it pressing down. "Yeah, it is," he admitted, his voice barely above a whisper. "It feels like I'm giving up on everything I've built. I mean, what will I be once I let go of this?"

Emily turned to face him fully, her gaze soft yet steady. "You're still you, Alex. Selling the agency doesn't change who you are. You're not defined by the agency, even if it feels that way right now. This place—it's something you created, yes, but you're so much more than the work you've done."

Her words grounded him, yet he still felt the ache. "Jordan kept saying I need to think about the legacy—about what happens to the team when I step back. I feel like I'm abandoning them. They've been with me through everything."

Emily squeezed his hand. "Do you really think you're abandoning them, or are you afraid they won't have what they need without you there?"

The question caught him off guard, and Alex's mind raced. Was his hesitation really about the team? Or was it a deeper fear—that his absence would somehow unravel all he'd done? Emily's words gave him clarity, a reminder that his leadership had never been about keeping control but about nurturing growth.

"You've taught them everything you know," Emily continued, her voice filled with a confidence he wished he could feel himself. "You created a culture that will carry on without you. Selling doesn't erase any of that. It proves you did it right—that they're ready to thrive on their own."

He nodded slowly, Emily's perspective helping him see the sale from a new angle. It wasn't about him stepping away; it was about the agency's evolution. "Jordan said something similar. He told me I have to let go of the need to hold everything together and start trusting in what I've built."

Emily smiled, her eyes warm. "You've done more than you realize, Alex. And maybe it's time for you to build something new for yourself, too. This doesn't have to be the end—it can be a new beginning."

In that moment, Alex felt a shift, like the tension that had held him in place was finally loosening. Emily's words had given him permission to embrace the future without fearing the past would be forgotten. He had built a foundation, created a legacy, and now, perhaps, it was time to let it stand on its own.

He took a deep breath, his gaze meeting Emily's, finding the strength he needed. "You're right," he said, his voice filled with a renewed resolve. "This isn't the end. It's just... the next step. And I think I'm ready for it."

The Last Drive: Decision at the Edge

Alex paced around his house, barely hearing Jordan's words as the numbers and agreements were laid out in front of him. It felt like any other meeting, as though this was just another deal, another client to land. But it wasn't—it was the end of his agency as he knew it, and the weight of that realization settled like a stone in his stomach.

Jordan's voice broke through the haze. "Alex, this is it. You've done everything right, all the numbers are there, the team's on board, and it's exactly the exit we planned for. $18 Million will be in your bank account in a matter of hours. You're ready for this."

Alex listened to his mentor, the man who had been with him through every tough decision, every late-night panic and breakthrough. Jordan's calm was unwavering, reassuring in a way Alex desperately wanted to feel himself. "I know you're saying it's all there, but what if..." Alex hesitated, struggling to find the right words, "...what if I'm not ready to actually walk away?"

Jordan lowered his tone. "Look, I know this is hard. But you and I both know this isn't about the agency anymore. This is about you, Alex. You're tied to this place because it's been part of your identity. But you've built a team that's ready, a culture that's solid, and a future that's bigger than you alone can take it. It's time to let them fly."

Alex swallowed, his throat tight. "But what if they need me? What if it all falls apart the second I'm out of the picture?"

Jordan's expression softened, and he placed a reassuring hand on Alex's shoulder. "They're ready because of you. If you go through with this, they'll thrive because of everything you've built. And remember, you'll still be a part of this, just in a new way—less control, yes, but more freedom to lead in a different way, one that doesn't cost you every piece of yourself. But you have to trust them."

A long silence settled between them, the gravity of Jordan's words sinking in. Alex finally nodded to himself, though his heart was racing, his thoughts scattered. He forced a smile, but even he could feel how unsteady it was. "Thanks, Jordan. I... I'll see you there."

As he left the house, Alex's mind was spinning. The drive to ClearEdge felt surreal, like stepping into a version of himself he couldn't quite recognize. Memories flickered through his mind—of late nights spent building the agency from the ground up, of celebrating wins with the team, of failures that had nearly brought them to their knees but always led to greater resilience. He felt himself smiling despite the anxiety, the ache of the decision throbbing in his chest.

But as he neared ClearEdge's headquarters, his heart pounded faster, a voice deep inside him growing louder. What if this was a mistake? What if he was abandoning something that only he could protect? He reached for his phone, his fingers hovering over Emily's contact. One ring, and she picked up.

"Hey, babe," she greeted him, her voice immediately comforting.

"Hey," he replied, his voice tight. "I know we've talked about this a thousand times, but I just—I don't know, Em. I thought I was ready, but now I'm not so sure. Everything's in place, and the team's excited, but..." He trailed off, struggling to finish the thought.

She was quiet for a moment before answering. "What are you afraid of, Alex?"

He clenched his jaw, trying to put words to the feeling twisting inside him. "I'm afraid it'll all fall apart if I'm not there. That they'll forget why we built this in the first place, or that the new owners won't understand what makes this company so special. I'm afraid my name and my reputation will be ruined if it all goes to shit after I sell."

Emily's voice was gentle but firm. "Alex, you've built something incredible here. But you and I both know it's taken a lot out of you. And it's taken a lot out of us, too. You've poured everything into this agency, but maybe it's time to pour a little back into yourself. You're not abandoning them. You're finally getting rewarded for all your hard work."

Her words hung in the air, both a reassurance and a challenge. He wanted to believe her, wanted to let go of the fear that seemed to be clawing at his chest, but the doubts still swirled. "But what if I regret it? What if I can't... I don't know, move on?"

She took a deep breath. "Then you'll face it, like you've faced every other challenge. But remember, you'll have something more than you have now. You'll have freedom, and maybe a chance to see what else life has to offer beyond the agency."

Her words struck a nerve, and Alex felt a pang of realization—the idea of freedom, of life outside the agency. He wasn't sure who he'd be without the agency, without the endless demands, the late nights, the pressure to keep everyone moving forward.

"Thanks, Em," he said quietly. "I just needed to hear that."

"Of course," she replied. "And remember, whatever you choose, I'm with you."

They hung up, and Alex set the phone down, staring out at the ClearEdge building in the distance. A sleek, polished facade that stood in stark contrast to his agency's raw, homegrown vibe. He'd poured everything into that place—his time, his energy, his vision. He'd watched it grow from a scrappy idea into something real, something that mattered. And now, he was about to walk away.

Or was he?

He tightened his grip on the steering wheel, his mind racing. It wasn't too late. He could turn the car around, head back to the office, call the team, and tell them the sale was off. They'd understand, wouldn't they? They'd welcome him back, and things would go back to normal.

But deep down, he knew that wasn't true. Things wouldn't go back to the way they'd been. He'd changed, the team had changed, the agency had grown beyond what one person could carry alone. And yet, the thought of stepping into that boardroom and signing away his creation felt impossible.

He sat there, the seconds ticking by, his mind replaying every decision, every sacrifice, every win, and every loss. What if this was a mistake? What if he was giving up too much? He felt the weight of the decision pressing down on him, suffocating and freeing at the same time.

In a final moment of clarity, he let go of the steering wheel, taking a deep breath, feeling the stillness settle around him. It was his choice, and his alone.

He opened the car door, stepped out, and closed it softly behind him. With each step toward the building, the future loomed larger, a shadowed possibility that was both exhilarating and terrifying.

At the door, he paused, his hand on the handle, his heart racing, the silence around him thick and heavy. This was it.

And in that suspended moment, as he stood on the edge of the decision that would define everything he'd built and everything he'd become, he hesitated—

The End

.

And What Might Your Marketing Agency Need for Growth?

If you're struggling to scale your agency and feel like you're running in circles, then you're not alone. Many agency owners face roadblocks when it comes to growth, often unsure where to start.

But what if you had a proven checklist that guides you through every critical step of scaling?
Inside this chapter, our character Alex talks about the very same Agency Growth Checklists that help him break free from the daily grind and build a business that practically runs itself.

Want to see the exact checklist that took him from chaos to control?
With actionable steps on everything from optimizing client fulfillment to refining team operations and boosting profit margins, this checklist can be yours.

Visit **www.factsnotfeelingsbook.com/checklists** and download all of our checklists for free and get the structure your agency needs to thrive.

Chapter 8: Lessons Learned
A Roadmap to Success

Hey my friend,

If you've made it here, first of all—thank you. Secondly, I hope you're not pissed at me for not giving you the ending you had hoped for in this book. And lastly, I hope you're feeling a little less alone. Because if there's one thing I wish I'd known earlier, it's that this journey isn't easy, and it's surely not something you have to do on your own.

When I look back on everything I've learned running an agency, I feel this urge to go back in time and give myself a good shake. I think about those long nights spent staring at the ceiling, those days of pulling out fire after fire, and, honestly, all the moments when I let my gut call the shots instead of letting facts lead me. I had no idea back then how much easier things could've been if I'd stopped running on pure adrenaline.

So that's why I wanted to write this letter, friend-to-friend, no fluff, no bull. This is about the stuff I wish someone had sat me down and told me years ago—the truths I stumbled into the hard way. I hope these words save you from some of those sleepless nights, the gut-churning worry that you're holding on by a thread, or that creeping fear that maybe, just maybe, you're in over your head.

1. Don't Try to Be the Hero of Every Problem

Here's the thing: being the "hero" feels good. It makes you feel needed, important, maybe even irreplaceable. But the need to save the day is a trap, and it's one I fell into hard. I wanted to be the guy who could fix anything, solve every problem, and jump into the middle of every crisis. It gave me this rush, like I was the heart and soul of my agency. But looking back, I see that by doing that, I was actually becoming the bottleneck. I was everywhere and nowhere at the same time, holding the agency together with duct tape and caffeine.

Your team doesn't need a hero; they need a leader. They need someone who believes in them, not someone who constantly steps in to take over. And that means giving up control, letting them step up—even if that means things won't be perfect. Trust me, letting go of that control is a lot harder than it sounds, but it's also the most freeing thing you'll ever do. It took me way too long to figure that out, and by the time I did, I had burned myself out.

If I could do it over, I'd remind myself that being the hero isn't sustainable. Being a leader is. And there's a huge difference.

2. Facts, Not Feelings: The Hardest and Best Lesson

If there's a mantra that's gotten me through the hardest

times, it's this: Facts, Not Feelings. It's easy to get wrapped up in the highs and lows, to feel the rush of a new deal or the sting of a lost client. And sure, those feelings matter. They keep us human. But they can't run your business. Feelings are fickle. Facts? Facts keep you steady when the storm hits.

One of the hardest days I had was realizing that "going with my gut" wasn't enough. I'd built this agency on instinct, and it worked at first. But as we grew, I needed something more solid to stand on. Numbers don't lie. They're there to give you clarity, direction, and a way to measure success that doesn't shift with every mood swing or gut reaction.

And don't get me wrong—it's not that feelings aren't important. They're just not everything. I learned that the hard way, and I wouldn't wish that lesson on anyone. So, if you're ever in doubt, if you're unsure whether to go left or right, start with the facts. They'll never lead you astray.

3. People Aren't Mind Readers

I can't count the number of times I assumed my team "just knew" what I wanted. It was like I expected them to be able to read my mind or know the endgame without me telling them. But the truth is, they didn't. And that was on me.

I had to learn that setting clear expectations isn't just helpful—it's everything. If you're frustrated because things aren't getting done the way you want, take a moment and ask yourself: have I actually given my team a clear roadmap? Have I been honest with them about what success looks like?

For a long time, I thought that if I hired smart people, they'd just figure it out. But that's not leadership. That's laziness. When you're clear, your team can thrive without constantly looking over their shoulders for your approval or trying to guess what you're thinking. And that's when you know you've built something real.

If I could go back, I'd tell myself to stop expecting people to "just know" and start vcommunicating like a real leader. Clarity isn't just kind—it's essential.

4. Don't Run from the Numbers (Especially When They're Ugly)

Here's a confession: I used to avoid looking at our financials. For years, I ran my agency like a kid with a piggy bank. I'd glance at what was left over at the end of the month and call it "business." But that's not business. That's survival. And it's a dangerous way to operate if you want to grow.

There's a certain fear that comes with looking at the numbers, especially when they're not pretty. I used to dread opening the books because I didn't want to face the reality of what I might see. But here's the thing: numbers are just information. They're not a reflection of your worth; they're a reflection of your strategy. When you finally look at the facts, you get to make decisions that put you in control.

I learned the hard way that if you're not willing to get

friendly with your numbers, you'll be in the dark. Gross Margins, EBITDA, cost of delivery—all of it. If you don't know what these mean or where you stand with them, get comfortable fast. Because the numbers are your map, your playbook. And the better you understand them, the more you'll know where to steer.

5. You Don't Need to Be Everything to Everyone

I used to think that in order to succeed, I had to offer everything. Whatever a client wanted, we'd find a way to do it, no matter how stretched we got. But in the end, all I did was create chaos.

Know what you're good at and do that. Don't try to please everyone. Not every client is a good fit, and that's okay. The minute I stopped trying to be everything to everyone, the agency felt like it was finally breathing again. Suddenly, we had time, focus, and a team that was doing work they actually enjoyed.

Focus on your strengths. Play to win, not just to survive. When I finally trimmed our service list down to what we did best, the results spoke for themselves. Less chaos, better margins, and a team that actually enjoyed their work. It wasn't easy, but it was the right call.

6. Stop Trying to Do It All

I used to think nobody could do it like me. And maybe that was true, but that didn't mean I should do it all. I was exhausted, stretched so thin that I couldn't give anything my full attention. It wasn't until I finally let go and started trusting my team that I realized how much I'd been holding everyone back—including myself.

If you want your agency to grow, you've got to step out of the way. Hire people who are better than you at specific things, give them the tools they need, and get out of their way. It's not about losing control; it's about building a team that doesn't need you hovering over their shoulders. And let me tell you, the freedom that comes from that is worth every bit of discomfort you feel while you're letting go.

7. Embrace the Fact That You'll Fail

Failure is uncomfortable, and I won't sugarcoat that. But it's also inevitable, and sometimes it's the best teacher you'll ever have. I learned more from my failures than I ever did from my successes. Each one taught me something I needed to know, something I couldn't have learned any other way.

Don't be afraid to fail. Don't try to dodge it. Every single setback is a lesson that moves you closer to success. The only time failure is a problem is when you let it stop you from moving forward.

8. Find Your "Why" and Hang On Tight

Running a business will test you. There will be days when you wonder if it's worth it, when it feels like everything you've built could come crashing down. That's when you need to hold tight to your "why"—the reason you started this journey in the first place.

For me, it was about building something I could be proud of, something that would make a difference, even if it was just in the lives of the people I worked with. That "why" kept me going when I wanted to give up. If you haven't found yours yet, find it. It's what will keep you grounded when things get tough.

9. Facts, Not Feelings, Will Set You Free

At the end of the day, if there's one thing that's made the biggest difference for me, it's this commitment to Facts, Not Feelings. Running an agency—or any business—is a rollercoaster. There are moments of pure excitement, and then there are days when it feels like everything is going wrong. But letting the facts, not your feelings, guide you is like having a compass that keeps you steady through it all.

Facts don't care if you're stressed or if you're on top of the world. They don't change. They're your anchor. The journey of running an agency isn't easy, but if you stay rooted in the facts, you'll be able to weather whatever comes your way.

So take this advice, make it your own, and remember that while feelings can inspire you, it's the facts that will keep you moving forward.

Your Friend,